D0931282

# Immobilized Enzymes

# Immobilized Enzymes

Author:
**Oskar Zaborsky, Ph.D.**
Esso Research and Engineering Company
Linden, New Jersey

published by:

**A DIVISION OF**
THE **CHEMICAL RUBBER** CO.
**18901 Cranwood Parkway • Cleveland, Ohio 44128**

This book represents information obtained from authentic and highly regarded sources. Reprinted material is quoted with permission, and sources are indicated. A wide variety of references are listed. Every reasonable effort has been made to give reliable data and information, but the author and the publisher cannot assume responsibility for the validity of all materials or for the consequences of their use.



Third printing September 1974

International Standard Book Number 0-87819-016-3

Library of Congress Card Number 72-95697

Division of THE CHEMICAL RUBBER CO.

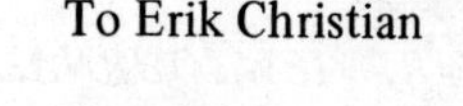

## THE AUTHOR

**Oskar R. Zaborsky** is a Senior Research Chemist at the Esso Research and Engineering Company in Linden, New Jersey, and has been a member of the Biosciences Group of the Corporate Research Laboratories since 1969.

Dr. Zaborsky attended the Philadelphia College of Pharmacy and Science (B.Sc. in Chemistry, 1964), the University of Chicago (Ph.D. in Chemistry, 1968), and Harvard University (postdoctoral fellow, 1968-1969). He earned his Ph.D. degree with Professor Emil T. Kaiser, studying the mechanism of hydrolysis of esters of sulfur-containing acids. His postdoctoral study was with Professor Elias J. Corey and involved model enzyme systems and mechanistic aspects of squalene cyclization. Both the studies at the University of Chicago and Harvard University were supported by National Institutes of Health Fellowships.

Dr. Zaborsky's present interests are the chemical modification of proteins and the immobilization of enzymes. He is involved in the development of new methods of immobilization and in the chemical and physical characterization of enzyme-polymer conjugates.

Dr. Zaborsky is a member of the American Chemical Society and the American Association for the Advancement of Science.

# PREFACE

Enzyme immobilization is a rapidly advancing scientific endeavor. At the moment, extraordinary activity exists in academic and industrial laboratories throughout the world. Government sponsored programs are also helping to create an enthusiastic climate, especially in Great Britain and the U.S. Publications on all aspects of immobilized enzymes are appearing in different journals, nearly a hundred articles being published during the last two years on just the polymer-bound type of immobilized enzyme. Such a publication rate obviously is indicative of a high interest in the subject, but even this volume may be just the "tip of the iceberg." Without question, a tremendous amount of research, conducted for years in a number of industrial laboratories, has not yet been disclosed. It is unfortunate that this material could not be included in this book. The work described in subsequent chapters should be considered only as a rough guide to the total quantity of research that has been conducted to date.

There are several reasons for writing this book. Although fine reviews have appeared recently on the subject of immobilized enzymes, no totally comprehensive review has yet been written. Katchalski, Goldstein, Silman, Mosbach, Melrose, and Kay have published excellent reviews, but they and others have limited their discussions to the classical methods for immobilizing enzymes.[1-34] They did not discuss the newer methods, nor did they present the total available literature. In addition, little was presented in a comparison of the individual methods with regard to their advantages and disadvantages. It is my hope that some of these shortcomings of previous reviews are corrected. This review, perhaps even more so than previous ones, is "method" oriented.

It is a formidable task to keep abreast of the overwhelming number of publications on immobilized enzymes which are appearing currently in all types of periodicals and to obtain copies of the original papers in a reasonable time. An added difficulty in "keeping up" with the literature is the fact that there are no accepted or approved rules of nomenclature for dealing with immobilized enzymes. Many terms are used in describing localized enzymes, and some of them, such as "insolubilization" and "immobilization," are frequently misused. Every attempt has been made to be as thorough as possible in covering the primary literature, and it is hoped that every article up to and including December 1971 has been incorporated in some form in the text or tables or is listed in the bibliography. An overt attempt was made to include only material from papers and certain patents already published. References not numbered in the bibliography are government reports and additional patents. These were not incorporated into the text or tables because much of what is discussed in them, especially the patents, has been reported by the same authors in the numbered references. Unpublished results from personal communications are included only if these are germane to the discussion. They have been conscientiously restricted to a bare minimum and are used only in the discussions of more recent developments.

This book is intended to be a fairly comprehensive treatise on immobilized enzymes and hopefully to present a complete summary of the state-of-the-art of this subject. It deals only with immobilized *enzymes* and not with proteins in general. Excellent reviews have been written on certain aspects of general protein immobilizations and references to these are found in Chapter 11, which also presents a brief summary of the potential of this technique in other scientific areas. At the very least, this book should give the reader a knowledge of the enzymes that have been immobilized and the methods that have been used. Tables of immobilized enzymes are arranged according to code numbers established by the IUB Commission on Enzyme Nomenclature. In the text, I have attempted primarily to present general observations and conclusions rather than to give a complete listing of all examples illustrating a particular aspect under discussion.

The quality of the work described by the hundreds of individuals mentioned in the bibliography has ranged from excellent to poor. Often, the properties of an immobilized enzyme were not compared with those of its soluble, unmodified counterpart – a fact that is sometimes reflected in my discussions. In order to limit the length of this review and to present only the more favorable, positive, and noteworthy aspects of these works, I have refrained from direct criticism of individual publications. Remarks made in the review that could be construed as critical generally should be considered more as personal preferences or possibilities for further investigation. The reader is, however, strongly urged to consult the orginal literature on critical aspects in order to judge for himself the merits and limitations of a particular work.

**Acknowledgments**

It is my pleasure to acknowledge the help of a number of people. I sincerely thank my colleagues in the Biosciences Group of the Corporate Research Laboratories – Drs. Allen I. Laskin, Sheldon W. May, and Robert D. Schwartz – for reading the manuscript and for many constructive comments. I especially thank Dr. Laskin for giving his time and support and for providing a conducive working environment. I also express my gratitude to Drs. Norman N. Li, Raam R. Mohan, and Peter R. Rony, Miss Jacqueline Ogletree, and Mr. Rex Sherwood for their help.

Many other people have contributed their time and effort – Misses Carol Bagwell, Kathi Hudock, Susan Koch, and Joan Donovan, who helped to type and correct the manuscript. I am especially grateful for the conscientious and untiring help of Miss Evelyn Petercsak for typing the final draft and for handling many items associated with the book, and Miss Marianne Morowski for typing a large part of the early draft. The staff of the Library, Photographic, and Forms and Graphics was very helpful. Mr. Raymond Moritz and Mr. Loenda Mier, Miss Betty McDevitt, and Mrs. Viola Boruszkowski provided guidance and prepared many of the illustrations.

In addition, I express my deep gratitude to Dr. Alan Schriesheim, Director of the Chemical Sciences Laboratory of the Corporate Research Laboratories, for providing the necessary aid to pursue this venture.

Last, I wish to thank most sincerely my wife, Marcia Lee, for typing early parts of the manuscript and for her unending patience during the months of writing.

Oskar R. Zaborsky
Linden,New Jersey
December 1972

# TABLE OF CONTENTS

Chapter 1

# INTRODUCTION

## What Is an Immobilized Enzyme?

The term "immobilization," as defined here and used throughout this book, means the physical confinement or localization of enzyme molecules during a *continuous* catalytic process. It is largely an operational definition intended to encompass a host of distinct and varied methods that have been used to achieve this aim. The localization of enzyme molecules can be brought about by such diverse means as their covalent attachment to water-insoluble functionalized polymers, their adsorption to water-insoluble inorganic or organic supports, their entrapment within gel matrices or semipermeable microcapsules, and their containment within semipermeable membrane-dependent devices. The word "continuous" is emphasized in order to limit the definition to a reasonable extent. Although an enzyme contained in a beaker or vial could be thought of as an "immobilized enzyme," it is not considered to be that in the context of the definition because the enzyme cannot be employed efficiently in a continuous manner. That is, an enzyme in a beaker or vial cannot be used readily as a continuous enzyme reactor in which a substrate is added and a product is removed continuously. On the contrary, an enzyme contained in an ultrafiltration cell or hollow fiber *is* considered to be "immobilized" within the scope of the definition. In these cases the enzyme is likewise physically localized in a specific space (defined by the semipermeable membrane and the walls), but the membrane allows free passage of the product and permits a *continuous* operational mode.

As defined in this manner, the term "immobilization" would also have to include enzymes that are chemically or physically attached to various water-insoluble components of cells such as the endoplasmic reticulum, mitochondria, nuclear membrane, and cell membrane. Likewise, it would be appropriate to use the term in describing enzymes contained naturally within cells, for these "localized" enzymes can certainly be used in a continuous fashion as enzyme reactors. Although such enzymes are indeed "immobilized" within the limits of our definition, they are not considered in the discussion; only enzymes that are *synthetically* immobilized are considered.

Classically, the term "immobilized enzyme" has been used to describe an enzyme that has been chemically or physically attached to a water-insoluble matrix, polymerized into a water-insoluble gel, or entrapped within a water-insoluble gel matrix or water-insoluble microcapsule. The term "water-insoluble" is used here repeatedly not for the sake of being redundant but for emphasizing an important point. In all these cases, the localization of the enzyme was achieved by producing a water-insoluble, enzyme-containing material. Consequently, it was quite natural that the two terms, "immobilization" and "insolubilization," erroneously came to mean the same to many. They are by no means equivalent. Insolubilization is just one of a number of methods that can be used for the immobilization of an enzyme. Another common misconception is that enzymes entrapped within the lattice structure of polymers or within microcapsules are "insolubilized enzymes." This again is not correct. These enzymes are only physically entrapped within a localized region and still have considerable diffusional freedom and do retain their inherent "solution" characteristics. They are simply restricted in their movements to a microspace.

Other sets of terms that are frequently misused are "stabilization" and "immobilization" or "insolubilization." Stabilization refers to the process of enhancing the resistance of an enzyme's properties to certain environmental changes. Again, these two combinations of terms – stabilization and insolubilization and stabilization and immobilization – have been used interchangeably. An "insoluble or immobilized enzyme" has been described often as being concomitantly "more stable" than the soluble enzyme solely because of its insoluble nature. This is certainly not true universally. There are many examples known where the immobilized enzyme (including the water-insolubilized, functionalized-polymer type) has diminished stability relative to its native counterpart. This question of stability is discussed in greater detail in Chaper 3.

It is perhaps appropriate to point out here that the terms "immobilization" and "immobilized enzyme" could be extremely beneficial as "key" words for use in publications dealing with synthetically produced, localized enzymes. At the present time, the literature abounds with a host of different terms that are used to describe this enzyme localization. For example, in the case of the polymer-bound enzyme method, terms such as "covalently linked, chemically bound, covalent fixation, attached, matrix-bound, insolubilized, immobilized, bonded, and coupled" have been used frequently. This abundance of terms makes it extremely difficult to locate papers. The use of "immobilization" or a derivative of this word is recommended as the "key" word in the title or abstract of papers.

**Why Immobilize Enzymes?**

There are three majors reasons for immobilizing enzymes. Immobilized enzymes: (1) offer a considerable operational advantage over freely mobile enzymes, (2) may exhibit selectively altered chemical or physical properties, and (3) may serve as model systems for natural, in vivo, membrane-bound enzymes. Not all of the methods discussed have an equal probability of achieving all or most of these aims. General operational advantages of immobilized enzymes are reusability, possibility of batch or continuous operational modes, rapid termination of reactions, controlled product formation, greater variety of engineering designs for continuous processes, and possible greater efficiency in consecutive multistep reactions. Certain methods for immobilizing enzymes – the chemically based methods – also possess the distinct possibility of changing the chemical or physical properties of enzymes in addition to immobilizing them. Chemically based methods involve the chemical reactivity of functional groups of enzymes and because certain organic functional groups are eliminated and changed to groups of different structures and reactivities, the immobilized enzymes may exhibit profound changes in their properties. In addition, certain types of immobilized enzymes can serve as model systems for the many experimental and theoretical aspects of natural, membrane-bound enzymes. The water-insoluble enzyme derivatives obtained by both chemical and physical means have been proposed for such studies. Chapter 10 deals in detail with the applications of immobilized enzymes.

**Classification System of the Methods Used for Immobilizing Enzymes**

There are several ways of classifying immobilized enzymes, and the classification system described here is, like any other, highly subjective. Methods for immobilizing enzymes are divided into two broad classes – chemical and physical. Chemical methods of immobilization include any that involve the formation of at least *one* covalent (or partially covalent) bond between residues of an enzyme and a functionalized, water-insoluble polymer or between two or more enzyme molecules. In reality, more than one covalent bond between the reacting components is normally formed. Chemical methods are usually irreversible in that the original enzyme cannot be regenerated or recovered. However, irreversibility per se is not an inherent feature of chemical methods; it is only a commonly found feature due to the nature of the chemical reactions employed for immobilizations.

Physical methods include any that involve localizing an enzyme in any manner whatsoever which is not dependent on covalent bond formation. In this class, the immobilization of enzymes is dependent on the operation of certain physical forces (electrostatic interaction, formation of ionic bonds, protein-protein interaction, etc.), the entrapment within microcompartments, or the containment in special membrane-dependent devices. In principle, physical methods of immobilization are completely reversible. However, many specific examples of the individual methods do not follow idealized behavior and considerable irreversible bond formation occurs.

Table 1 gives the classification system and lists the various individual methods that are discussed in detail in subsequent chapters. The subdivision of the physical methods is especially arbitrary and has been retained more from a historical than from a purely physical basis. Further, in many specific examples of immobilization to be discussed, the exact nature of the forces responsible for the immobilization is itself either multivariant or not clearly categorizable. Although a particular example may be included in one category, the

TABLE 1

**Classification System of the Methods Employed for the Immobilization of Enzymes**

Chemical methods—(Covalent bond formation-dependent)
- Attachment of enzyme to water-insoluble, functionalized polymer
- Incorporation of enzyme into growing polymer chain
- Intermolecular crosslinking of enzyme with a multifunctional, low molecular weight reagent

Physical methods—(Noncovalent bond formation-dependent)
- Adsorption of enzyme onto water-insoluble matrix
- Entrapment of enzyme within water-insoluble gel matrix (lattice-entrapment)
- Entrapment of enzyme within permanent and nonpermanent semipermeable microcapsules
- Containment of enzyme within special semipermeable membrane-dependent devices

immobilization of the enzyme may actually involve several distinct processes (e.g., adsorption and entrapment or covalent linkage and adsorption).

Chapter 2

# COVALENT ATTACHMENT TO WATER–INSOLUBLE FUNCTIONALIZED POLYMERS

Chemical methods for the immobilization of enzymes are based on the formation of covalent bonds between enzyme molecules and functionalized polymers or other enzyme molecules. This chapter deals with the covalent attachment of enzymes to water-insoluble, functionalized polymers and with the preparation of enzyme derivatives via copolymerization reactions. It discusses the various methods employed and the nature of the supports used to achieve this type of immobilization. Chapter 3, in turn, presents a discussion of the chemical and physical properties of these water-insoluble, polymer-bound enzyme conjugates.

The covalent attachment of water-soluble enzyme molecules via nonessential amino acid residues to water-insoluble, functionalized supports is the most prevalent method for immobilizing enzymes. Much has been written on various aspects of the method, and it is the one that commonly comes to mind when referring to immobilized enzymes. The water-insoluble supports can be either organic or inorganic polymers. Other commonly used names for the support are "carrier," "matrix," or "polymer." Likewise, several terms are used to describe the covalent bond formation between the enzyme and support; terms such as "fixation, linkage, attachment, binding, bonding, coupling, and grafting" are employed frequently.

The first example of a *synthetically* produced, water-insoluble, polymer-bound enzyme derivative was described by Grubhofer and Schleith[35,36] in 1953. Several enzymes ($\alpha$-amylase, pepsin, ribonuclease, and carboxypeptidase) were immobilized on diazotized polyaminostyrene to give water-insoluble, enzymatically active, enzyme-polymer conjugates. These investigators also reported the immobilization of the four enzymes on a chlorinated resin (Amberlite® XE-64 previously modified with thionyl chloride) but unfortunately neglected to give full data. Although their work is commonly considered to be the first example of a *covalently* linked enzyme-polymer adduct, and it is no doubt the case, it should be pointed out that Grubhofer and Schleith did not establish unequivocally that their enzyme-polymer conjugates were solely covalently bonded. A considerable portion, if not all, of the immobilization could have been caused by physical adsorption and not by covalent bond formation.

The question of the relative contributions of adsorption and covalent linkage to the total bound enzyme on a support was discussed by Brandenberger[37,38] in 1955. Brandenberger found that catalase and lipase were as strongly bound to unmodified polyaminostyrene and Amberlite XE-97 (a carboxylic ion-exchange resin) as they were to diazotized polyaminostyrene and chlorinated (thionyl chloride treated) XE-97 resin. The enzyme-polymer conjugates were washed with saline, saline and water, and saline, water, and phosphate buffer solutions in order to remove any physically adsorbed enzyme, and in every case, the modified and unmodifed enzyme-polymer conjugates exhibited similar activities. Brandenberger thus cast suspicion on the earlier claims of Grubhofer and Schleith that the immobilization of enzymes on diazotized polyaminostyrene was due only to covalent bond formation between the enzymes and support. Brandenberger's own system for studying the relative contributions of adsorption and covalent linkage to the total bound protein of an immobilized enzyme consisted of catalase or lipase immobilized on polyaminostyrene modified by phosgene (i.e., *p*-isocyanatostyrene). His evidence for claiming *covalent bond* formation between the protein and the *p*-isocyanatostyrene was based on the lower amount of carbon dioxide liberated during the reaction of the functionalized polymer with an aqueous solution of either a protein or glycine compared with buffer or water alone. The chemistry behind his reasoning is given in Equations 1 and 2.

$$RNCO + R'NH_2 \longrightarrow RNHCONHR' \quad (1)$$

$$RNCO + HOH \longrightarrow RNHCO_2H \longrightarrow RNH_2 + CO_2 \quad (2)$$

Isocyanates react with primary amines to give substituted ureas. On the other hand, isocyanates react with water to give unstable carbamic acid derivatives which decompose to primary amines and carbon dioxide.

From these initial studies in the mid-fifties, the number of examples of this method for immobilizing enzymes has grown tremendously. Nearly a hundred papers have appeared on this mode of immobilization alone within the last two years. Enzymes have been covalently attached to organic and inorganic supports of both natural and synthetic origin. A complete list of the enzymes immobilized covalently on supports and through various functionalities is presented in Table 4 at the end of this chapter. A description of the major supports used and the chemistry employed in the activation of the polymers and in their coupling with enzymes follows.

**Methods for Preparing Water-Insoluble Enzyme-Polymer Conjugates**

The most commonly employed method for preparing synthetic enzyme-polymer conjugates consists of contacting an aqueous enzyme solution with a water-insoluble, functionalized reactive polymer. Other general variations are also possible. Water-insoluble enzyme-polymer conjugates can be produced by (1) the copolymerization of the enzyme with a reactive monomer and (2) by a combination of methods such as the simultaneous covalent bonding and entrapment of an enzyme. The general methods are given in Figure 1 in outline form on page 8.

The preparation of water-insoluble, enzyme-polymer conjugates from *preformed,* functionalized polymers can be divided further arbitrarily according to the conditions used to obtain the adduct, and Table 2 gives the variations commonly used to obtain such surface-bound, enzyme-polymer conjugates.

The simplest procedure for producing water-insoluble surface-bonded, enzyme-polymer conjugates shown in Table 2 consists of contacting an enzyme solution with a preformed reactive polymer. The polymers used are synthesized from monomeric units which contain the reactive functionality, and no activation is required. Examples of such reactive polymers are maleic anhydride-based copolymers, methacrylic acid anhydride-based copolymers, nitrated fluoroaryl methacrylic copolymers, and iodalkyl methacrylates. These supports are synthesized in nonaqueous solvents and decompose in water. Consequently, in an aqueous solution of an enzyme, competitive reactions between functional groups of the enzyme and water (or buffer) largely determine the eventual success of the method.

Other subdivisions listed in Table 2 are arbitrary in nature, but they are helpful in pointing out the gradation in stabilities of reactive functional groups of the polymers, the various modes of addition of enzyme and polymer, and the distinction in the use of multifunctional reagents for enzyme immobilization. Method B consists of activating the polymer through a chemical conversion of its functional groups followed by addition of the enzyme. Common examples of this type of immobilization are the diazotization of aminoaryl polymers and the cyanogen bromide activation of hydroxyl-containing polymers. Method C consists of the simultaneous addition of a polymer-activating reagent and the enzyme. An example of the method is the activation of carboxyl-containing polymers with water-soluble carbodiimides. The addition of the carbodiimide and enzyme is usually, but not always, performed simultaneously. It can also be done sequentially with first the carbodiimide being added to the polymer to give the reactive acylisourea intermediate which is filtered, washed, and then reacted with the enzyme. Preformed functionalized polymers can be activated also through the use of a low molecular weight, bi- or multifunctional reagent. The water-insoluble functionalized polymer is treated with a large excess of the reagent to give a derivatized polymer which has as its reactive functionality one part of the multifunctional reagent. Examples of such activated polymers are aminoethylcellulose (AE-cellulose) or polyacrylamide treated with glutaraldehyde and the celluloses treated with various triazinyl derivatives. Used in this manner, one part of the multifunctional reagent is employed for the introduction of the reactive group to the polymeric support and the other part is used for covalent coupling of the enzyme. Two groups of a low molecular weight bi- or multifunctional reagent can also be used to produce water-insoluble enzyme derivatives without the use of a preformed polymer. This type of immobilization is presented in Chapter 4.

**General Nature of the Enzyme to be Immobilized**

The immobilization of a water-soluble enzyme

TABLE 2

**Variations for Producing Water-Insoluble, Surface-Bonded, Enzyme-Polymer Conjugates from Preformed, Water-Insoluble Supports**

Reactions of the enzyme with:

A. Reactive polymer (no activation procedure involved) (e.g., maleic anhydride copolymers)

B. Polymer (activation procedure involved; activation involves the chemical conversion of a functional group of the polymer followed by the addition of the enzyme) (e.g., diazotization of polyaminostyrenes)

C. Polymer (activation procedure involved; activation involves the chemical conversion of a functional group of the polymer with a reagent in the presence of the enzyme) (e.g., carbodiimide coupling with carboxyl-containing polymers)

D. Polymer (activation procedure involved; activation involves the chemical conversion of a functional group of the polymer with a multifunctional, low molecular weight, reagent followed by the addition of the enzyme) (e.g., glutaraldehyde modification of amino-containing polymers)

by covalent attachment to a water-insoluble polymeric support should involve only functional groups of the enzyme that are not essential for catalysis and those, if reacted, which would not alter detrimentally any chemical or physical properties of the enzyme. Thus, the "active center" residues and functional groups needed for purposes such as allosteric control or subunit interaction should, if possible, be avoided. Amino acid residues of enzymes that have been either implicated or proven to be involved in the covalent bond formation with supports are lysine (also terminal $\alpha$-amino groups), arginine, histidine, tryptophan, tyrosine, cysteine, aspartic and glutamic acids (also terminal carboxyl groups), and serine.

Unsuccessful examples of enzyme immobilizations have been reported. Failure was caused by not achieving covalent coupling between the enzyme and the polymer, or more often, by the chemical inactivation of the enzyme. Specific examples of unsuccessful binding are alkaline phosphatase with cyanogen bromide-activated Sephadex® and with the azide of carboxymethylcellulose (CM-cellulose).[39] Examples of covalent binding of enzymes but with concomitant loss of activity are pyruvate kinase bound to diethylaminoethylcellulose (DEAE-cellulose) via cyanuric chloride,[40] trypsin and dextranase bound to the nitrated copolymer of methacrylic acid and methacrylic acid-*m*-fluoroanilide,[41] and $\gamma$-amylase bound to the isothiocyanato derivative of 3-(*p*-aminophenoxy)-2-hydroxypropylcellulose.[42] More examples of inactivation are found in Table 4.

To prevent, or at least partially eliminate, inactivation of the enzyme upon its covalent attachment to a water-insoluble support, several measures can be used. Those that achieve higher activities in the resulting immobilized enzyme by preventing reaction with the essential amino acid residues include the covalent attachment of (1) the enzyme in the presence of a competitive inhibitor or substrate, (2) a reversible covalently linked enzyme-inhibitor complex, (3) a chemically modified enzyme whose covalent linkage to the polymer is achieved by newly incorporated residues, and (4) a zymogen precursor. For example, the covalent attachment of chymotrypsin in the presence of the competitive inhibitor, $\beta$-phenylpropionic acid, with poly(iodoalkyl methacrylates) (Poliodals) led to catalytically more active water-insoluble enzyme derivatives.[43] On the other hand, the presence of *N*-acetyl-DL-tryptophan, another competitive inhibitor of chymotrypsin, during its coupling to cyanogen bromide-activated agarose had no significant effect on either the enzymic activity or stability of the resulting derivative.[44] In one example cited, the presence of a substrate had a deleterious effect on the physical nature of the enzyme-polymer derivative. Glassmeyer and Ogle[45] reported that the attachment of trypsin to an amino-containing polymer through the glutaraldehyde extension procedure in the presence of

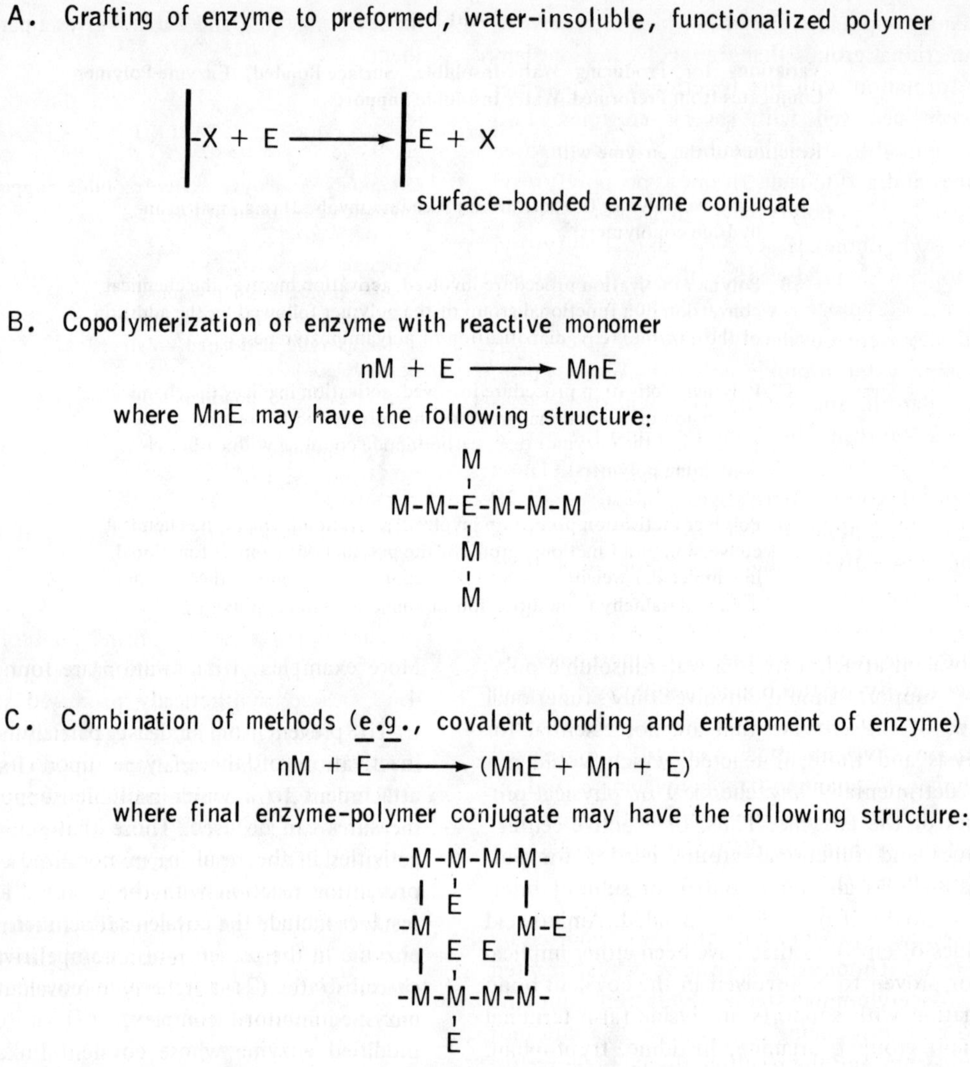

A. Grafting of enzyme to preformed, water-insoluble, functionalized polymer

$$|\text{-X} + \text{E} \longrightarrow |\text{-E} + \text{X}$$

surface-bonded enzyme conjugate

B. Copolymerization of enzyme with reactive monomer

$$n\text{M} + \text{E} \longrightarrow \text{MnE}$$

where MnE may have the following structure:

```
       M
       |
  M-M-E-M-M-M
       |
       M
       |
       M
```

C. Combination of methods (e.g., covalent bonding and entrapment of enzyme)

$$n\text{M} + \text{E} \longrightarrow (\text{MnE} + \text{Mn} + \text{E})$$

where final enzyme-polymer conjugate may have the following structure:

```
 -M-M-M-M-
  |  E   |
 -M     M-E
  |  E E  |
 -M-M-M-M-
     E
```

FIGURE 1. General variations for preparing synthetic, water-insoluble, enzyme-polymer conjugates.

α-*N*-benzoyl-L-arginine amide (BAA) led to insoluble products that were gummy and could not be pipetted.

The covalent attachment of an active site-modified enzyme derivative to a water-insoluble functionalized polymer has been achieved with several enzymes through their *p*-chloro- or *p*-hydroxymercuribenzoate derivatives. Enzymatically inactive *p*-mercuribenzoate derivatives of urease,[46] papain,[47,48] ficin,[49] and ATPase,[50,51] were covalently linked to various water-insoluble supports and then activated with sulfhydryl reagents or acid to their enzymatically active forms. With ficin, the activity after washing with cysteine was found to be similar to that of the unmodified enzyme. On the contrary, *p*-chloromercuribenzoate-modified urease exhibited storage stability (i.e., reactivation to its active form) superior to that of the unmodified enzyme. Similarly, the modified derivatives of papain and ATPase showed greater enzymic activity after reactivation. This was especially the case with ATPase; without first blocking the enzyme with *p*-hydroxymercuribenzoate, no activity was observed in the water-insoluble derivative obtained in the reaction of the enzyme with the azide of carboxymethylcellulose.

The incorporation of "auxiliary" groups

(pendant groups) into enzymes which then serve as the functional groups that are involved in covalent bond formation with the functionalized polymer has been achieved with several enzymes. Two types of modifications have been used with three enzymes and a zymogen. In one type, polytyrosyl trypsin,[52,53] polytyrosyl α-chymotrypsin,[53] polytyrosyl ribonuclease A,[53] and polytyrosyl prothrombin,[54] produced by the reaction of the enzyme or zymogen with *N*-carboxy-L-tyrosine anhydride, were covalently linked to diazonium-containing water-insoluble polymers. With trypsin, Glazer, Bar-Eli, and Katchalski[53] reported a 19.0 and 13.7% tyrosine incorporation into the parent molecule in two separate preparations. These polytyrosyl trypsin derivatives, indicative of 28 and 20 added moles of tyrosine per molecule of trypsin, respectively, and containing an average tyrosine chain length of about 2.5 (for the 13.7% preparation), retained full enzymic activity and substrate specificity. Further, a dramatic increase in the activity of the bound polytyrosyl trypsin compared with the unmodified enzyme was observed on covalent attachment to a diazotized copolymer of *p*-amino-DL-phenylalanine and L-leucine.[55] The reactions involved in the modification of the proteins are discussed later in the text (see Equation 37).

A second type of modification was reported by Epton and Thomas.[56] Trypsin was first reacted with *N*-acetylhomocysteine thiolactone for the purpose of introducing free sulfhydryl groups into the molecule, and the resulting thiolated derivative was then covalently attached to Enzacryl® Polythiol. The structure and chemistry of the Enzacryl supports are discussed elsewhere. Equations 3 and 4 give the essence of the method.

The covalent attachment of a zymogen to a water-insoluble polymer followed by its activation could be used to achieve greater enzymatic activity in the resulting enzyme-polymer conjugate. Covalently attached zymogens of trypsin, α-chymotrypsin, thrombin, and plasmin are included in Table 4, but they were not immobilized expressly for the purpose of increasing the activity of an enzyme.

## Functionalized Supports Used for Enzyme Immobilization

A host of widely different water-insoluble supports are employed for the covalent attachment of enzymes, and Table 3 gives a partial list of them.

TABLE 3

**Commonly Employed Water-Insoluble Supports for the Covalent Attachment of Enzymes**

Synthetic supports
- Acrylamide-based polymers
- Maleic anhydride-based polymers
- Methacrylic acid-based polymers
- Polypeptides
- Styrene-based polymers

Natural supports
- Agarose (Sepharose)
- Cellulose
- Dextran (Sephadex)
- Glass
- Starch

The supports can be organic or inorganic and range from materials found in nature to highly sophisticated, synthetically produced substances. In Table 4 are listed, under each enzyme, specific supports and the variations in activation procedures used to couple particular enzymes. This section deals first with some of the earlier, more commonly employed supports and/or methods, and then with more recent ones. The recent methods should give the reader a feeling for the present direction of research in this field. The chemistry discussed is fairly representative of all methods described to date.

### *Established Methods for the Covalent Attachment of Enzymes to Water-Insoluble Functionalized Supports*

#### ***Carboxymethylcellulose Azide***

One of the oldest and most frequently employed methods for the covalent linkage of an enzyme to a support is with the azide of CM-cellulose. Chemical reactions leading to the active support from cellulose and coupling with an enzyme are shown in Equations 5 and 6.

All transformations are simple and yields are relatively high. The hydrazide of CM-cellulose is commercially available, and the only activation step needed is treatment of the hydrazide with nitrous acid to form the acid azide. The coupling reaction is an acylation reaction involving predominantly the ε-amino group of lysine to form the amide linkage. Other amino acid residues implicated in covalent bonding are tyrosine, cysteine, and serine. The

$$\text{H}_2\text{N-Trypsin-NH}_2 + \underset{\text{N-Acetylhomocysteine thiolactone}}{\text{CH}_3\text{CONH-(thiolactone ring, S, C=O)}} \longrightarrow \text{CH}_3\text{CONH-CH(CH}_2\text{CH}_2\text{SH)-CONH-Trypsin-NHCOCH(NHCOCH}_3\text{)CH}_2\text{CH}_2\text{SH} \quad (3)$$

N-Acetylhomocysteine thiolactone

$$-\text{CH}_2\text{-CH}(\text{CO-NH-CH(CO}_2\text{H)CH}_2\text{SH})- + \text{Trypsin-NHCOCH(NHCOCH}_3\text{)CH}_2\text{CH}_2\text{SH} \xrightarrow{\text{K}_3\text{Fe(CN)}_6}$$

Enzacryl Polythiol

$$-\text{CH}_2\text{-CH}(\text{CO-NH-CH(CO}_2\text{H)-CH}_2\text{S-SCH}_2\text{CH}_2\text{CH(NHCOCH}_3\text{)CONH-Trypsin})- \quad (4)$$

Activation:

$$\text{Cellulose-OH} \xrightarrow[\text{NaOH}]{\text{ClCH}_2\text{CO}_2\text{H}} \underset{\text{CM-cellulose}}{\text{Cellulose-O-CH}_2\text{CO}_2\text{H}} \xrightarrow[\text{HCl}]{\text{CH}_3\text{OH}} \text{Cellulose-O-CH}_2\text{CO}_2\text{CH}_3 \xrightarrow{\text{NH}_2\text{NH}_2}$$

$$\text{Cellulose-O-CH}_2\text{CONHNH}_2 \xrightarrow[\text{HCl}]{\text{NaNO}_2} \text{Cellulose-O-CH}_2\text{CON}_3 \quad (5)$$

Coupling:

$$\text{Cellulose-O-CH}_2\text{CON}_3 + \text{E-NH}_2 \longrightarrow \text{Cellulose-O-CH}_2\text{CONH-E} \quad (6)$$

method was first applied to the enzymes, $\alpha$-chymotrypsin and trypsin, in 1961 by Mitz and Summaria.[57]

### *Polyaminostyrenes*

Another support used early for enzyme immobilizations is poly-*p*-aminostyrene activated either with nitrous acid, phosgene, or thiophosgene to give the corresponding diazonium, isocyanato, and isothiocyanato derivatives, respectively. Poly-*p*-aminostyrene is commercially available and its conversion to the active deriva-

Activation

(7) $NaNO_2$ / HCl → $-CH-CH_2-$ with $N_2^+Cl^-$

(8) $Cl-C(=O)-Cl$ → $-CH-CH_2-$ with NCO

(9) $Cl-C(=S)-Cl$ → $-CH-CH_2-$ with NCS

Starting polymer: $-CH-CH_2-$ with $NH_2$

tives is easily accomplished (Equations 7 to 9). Grubhofer and Schleith[35,36] first coupled α-amylase, pepsin, ribonuclease, and carboxypeptidase to diazotized poly-*p*-aminostyrene, and Brandenberger attached catalase and lipase[37,38] to the corresponding isocyanato derivative. Although the *p*-isothiocyanato derivative was employed for protein (albumin) immobilization by Manecke and Singer,[58] it has not yet been used for enzyme insolubilization. Various poly-*m*-aminostyrenes have been employed also for enzyme immobilization (see Table 4). The reactions of a diazonium salt with proteins are alkylation reactions involving predominantly the tyrosine and histidine residues, but others (lysine, arginine, and cysteine) are implicated. Isocyanato and isothiocyanato groups react predominantly with the primary amino groups of lysyl residues (ε-amino groups) to form substituted ureas and thioureas, respectively. Arginine has also been implicated in the reaction with isothiocyanato derivatives (Equations 10 to 12).

### Copolymers of p-Aminophenylalanine

The use of water-insoluble polypeptides as enzyme immobilization supports is well established and they have been used by many investigators and for many different enzymes. The copolymer most commonly employed is that of L-leucine and *p*-amino-DL-phenylalanine. The support is produced by the copolymerization of α,*N*-carboxy-*p*-amino, *N*-benzyloxycarbonyl-DL-phenylalanine anhydride and *N*-carboxy-L-leucine

Coupling:

(10) $-CH-CH_2-$ with $N_2^+Cl^-$ + E-Tyr → $-CH-CH_2-$ with $N=N$–(aryl with OH, $CH_2$-E)

(11) $-CH-CH_2-$ with NCO + E-$NH_2$ → $-CH-CH_2-$ with NHCONH-E

(12) $-CH-CH_2-$ with NCS + E-$NH_2$ → $-CH-CH_2-$ with NHCSNH-E

Preparation and Activation:

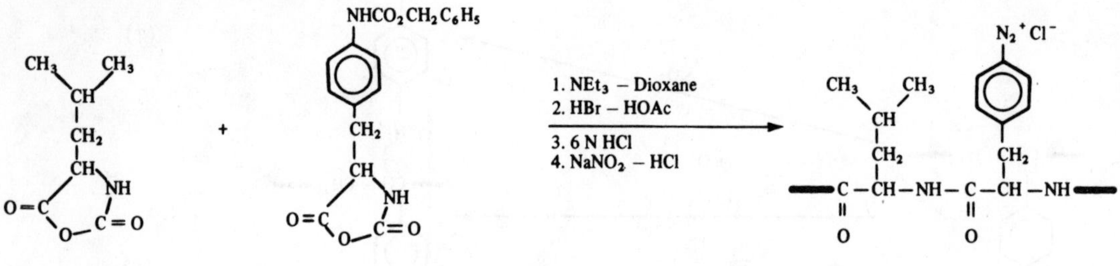

$NHCO_2CH_2C_6H_5$

1. $NEt_3$ – Dioxane
2. HBr – HOAc
3. 6 N HCl
4. $NaNO_2$ – HCl

$N_2^+Cl^-$ (13)

*p*-Amino-DL-Phe – L-Leu copolymer

Coupling:

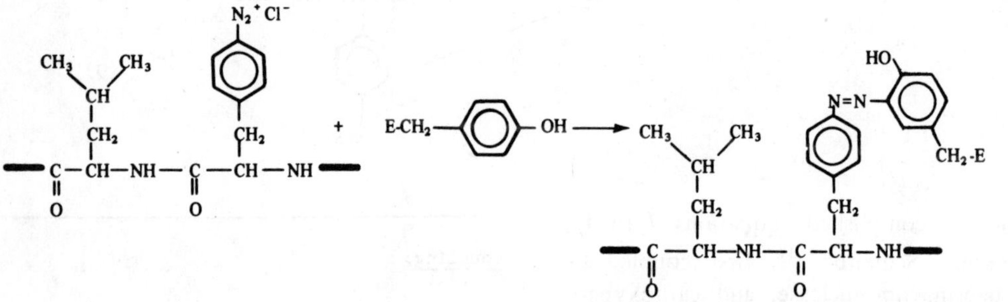

$N_2^+Cl^-$

E-$CH_2$–⟨C₆H₄⟩–OH (14)

anhydride in dioxane using triethylamine as an initiator,[52] followed by treatment of the resulting copolymer with hydrogen bromide in glacial acetic acid and by 6 *N* HCl. The copolymer is activated with nitrous acid to give the diazonium salt. Equations 13 and 14 give the reactions involved in its preparation, activation, and eventual coupling with enzymes.

### *Aminoaryl-Containing Celluloses*

A series of aminoaryl-containing celluloses has been employed for enzyme immobilization via the diazotization procedure. The most frequently employed aminoaryl derivatives of cellulose are the *p*- and *m*-aminobenzyl ethers, *p*-aminobenzoyl ester, and 3-(*p*-aminophenoxy)-2-hydroxypropyl ether. The last derivative of cellulose mentioned has been converted also to its isothiocyanato derivative. Cellulose ethers and esters can be prepared from cellulose and the appropriate activated alkyl halide or acid chloride (see Equations 15 and 16).

### *Aminoalkyl and Aminoaryl Derivatives of Glass*

The use of glass as a support for the covalent attachment of enzymes was first reported by Weetall[59,60] in 1969. Glass is claimed to be a superior support for enzyme immobilization because of its structural stability, inertness to microbial attack, and its easily shaped character. Most glass used for covalent enzyme immobilization is the Corning porous 96% silica glass. Both the particle size (usually 40 to 80 mesh) and mean pore diameter (anywhere from approximately 200 to 1500 Å) can be controlled rigidly. Although a number of organic functional groups can be grafted onto the surface of glass, all involve first the preparation of the 3-aminopropyl silane derivative. The preparation of the "aminoalkyl" derivative and its subsequent transformation into the "aminoaryl" derivative are shown in Equation 17.

### *s-Triazinyl Derivatives of Cellulose*

A commonly employed method for the immobilization of enzymes on various types and shapes of celluloses is through the activation of the polysaccharides with cyanuric chloride or some of its dichloro derivatives.[61–66] 2-Amino-4,6-dichloro-*s*-triazine has been the dichloro derivative of choice, but others used are the 2-carboxymethylamino and 2-carboxymethyloxy derivatives and complex structured dyes such as Procian® Brilliant Orange MGS. Activation of the celluloses and coupling of enzymes is easily achieved (see

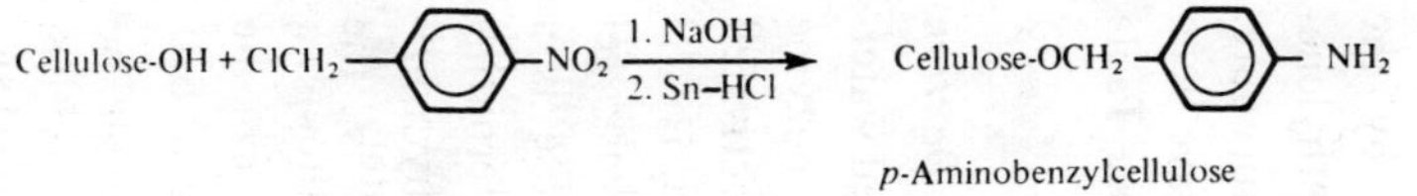

$$\text{Cellulose-OH} + ClCH_2-C_6H_4-NO_2 \xrightarrow[\text{2. Sn–HCl}]{\text{1. NaOH}} \text{Cellulose-OCH}_2-C_6H_4-NH_2 \quad (15)$$

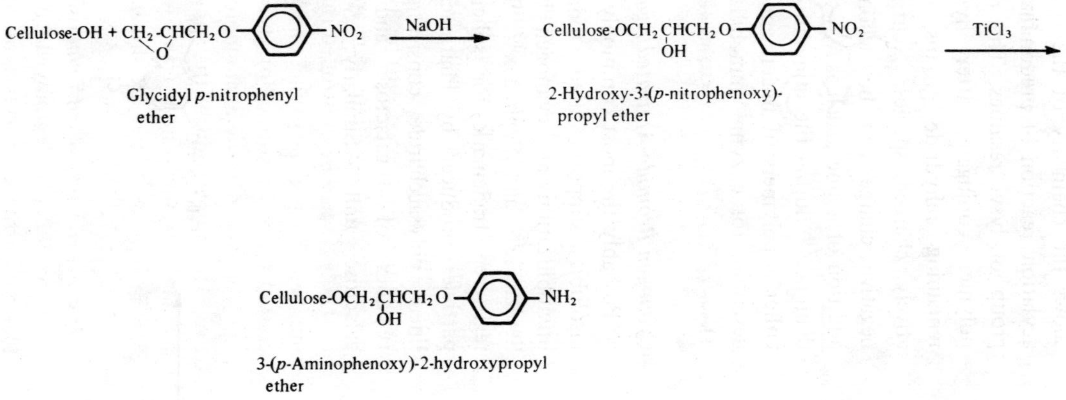

*p*-Aminobenzylcellulose

$$\text{Cellulose-OH} + \underset{\diagdown O \diagup}{CH_2\text{-}CHCH_2}O-C_6H_4-NO_2 \xrightarrow{\text{NaOH}} \text{Cellulose-OCH}_2\underset{OH}{CH}CH_2O-C_6H_4-NO_2 \xrightarrow{TiCl_3}$$

Glycidyl *p*-nitrophenyl ether

2-Hydroxy-3-(*p*-nitrophenoxy)-propyl ether

$$\text{Cellulose-OCH}_2\underset{OH}{CH}CH_2O-C_6H_4-NH_2 \quad (16)$$

3-(*p*-Aminophenoxy)-2-hydroxypropyl ether

Preparation:

$$-O-\underset{O}{\overset{O}{Si}}-OH + EtO-\underset{OEt}{\overset{OEt}{Si}}-CH_2CH_2CH_2NH_2 \longrightarrow -O-\underset{O}{\overset{O}{Si}}-O-\underset{O}{\overset{O}{Si}}-CH_2CH_2CH_2NH_2 \xrightarrow[\text{2. Sodium dithionite}]{\text{1. }p\text{-Nitrobenzoyl chloride}}$$

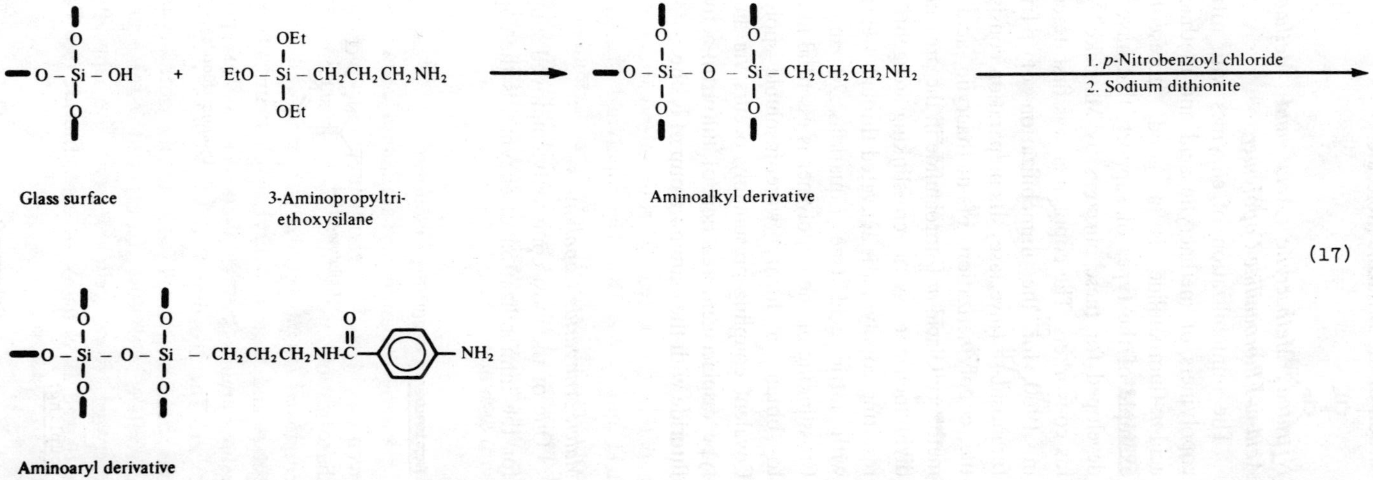

Glass surface

3-Aminopropyltriethoxysilane

Aminoalkyl derivative

$$-O-\underset{O}{\overset{O}{Si}}-O-\underset{O}{\overset{O}{Si}}-CH_2CH_2CH_2NH-\overset{O}{\overset{\|}{C}}-C_6H_4-NH_2 \quad (17)$$

Aminoaryl derivative

Equations 18 and 19). The coupling reaction is an alkylation reaction involving the primary amino groups of the enzyme (again, predominantly the $\alpha$- and $\epsilon$-amino groups of lysine) with the activated carbon of the *s*-triazine molecule.

### *Nitrated Methacrylic Acid and Methacrylic Acid-m-Fluoroanilide Copolymer*

The immobilization of enzymes with nitrated copolymers of methacrylic acid and methacrylic acid-*m*-fluoroanilide is a good representative example of the types of supports that have been developed for these purposes by Manecke[67] and his co-workers. The copolymer was first described in 1960 for the immobilization of $\beta$-fructofuranosidase (invertase). Its preparation consists of the copolymerization of methacrylic acid and methacrylic acid-*m*-fluoroanilide in the presence of divinylbenzene as a crosslinking reagent. The resulting copolymer is activated through treatment with nitric acid (see Equations 20 and 21). Crosslinking of the copolymer is essential in order to obtain a highly water-insoluble polymer. Covalent coupling presumably occurs via an $S_N2$ type displacement reaction of the nitro-activated fluoride with the $\epsilon$-amino group of lysine.

### *Maleic Anhydride Copolymers*

One of the most commonly employed supports for the immobilization of enzymes is the copolymer of maleic anhydride and ethylene. The use of this support was first reported by Levin et al.[68,69] in 1964. It is a commercially available copolymer, and immobilization of the enzyme consists of the addition of the protein to the anhydride-containing support. A water-soluble diamine (e.g., 1,6-diaminohexane) is added to enhance the crosslinking of polymer chains in order to produce a highly water-insoluble enzyme resin. Equation 22 shows the chemistry of the coupling step – an acylation reaction of predominantly the $\epsilon$-amino group of lysyl residues. The resulting enzyme-polymer conjugate, after hydrolysis of any remaining anhydride groups, is a highly negatively charged species. Neutralization of the negative charge can be accomplished by the addition of amines, such as *N,N*-dimethylethylenediamine,[70] during the immobilization procedure. Other copolymers of maleic anhydride have been described for enzyme immobilization (see Table 4).

### *Cyanogen Bromide-Activated Agarose and Dextran*

Probably the most commonly employed water-insoluble supports used today for enzyme immobilization are cyanogen bromide-activated Sepharose® and Sephadex®. Sepharose is the registered trademark for spherical *agarose* gel particles produced by Pharmacia. Agarose is a linear polysaccharide consisting of alternating residues of D-galactose and 3,6-anhydro-D-galactose units. Similarly, Sephadex is the

Activation:

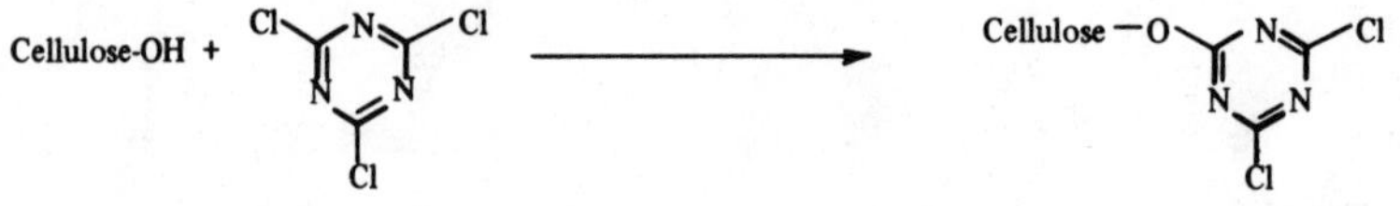

Cellulose-OH + Cyanuric chloride (2,4,6-trichloro-s-triazine) ⟶ Cellulose—O–(4,6-dichloro-s-triazin-2-yl) (18)

Coupling:

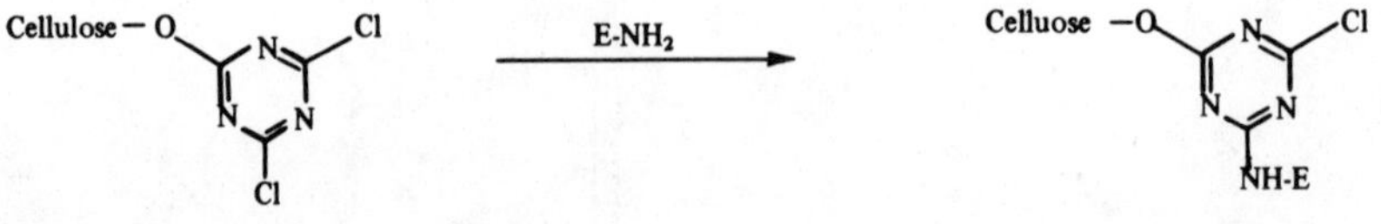

Cellulose—O–(4,6-dichloro-s-triazin-2-yl) $\xrightarrow{\text{E-NH}_2}$ Celluose —O–(4-chloro-6-NH-E-s-triazin-2-yl) (19)

**Preparation:**

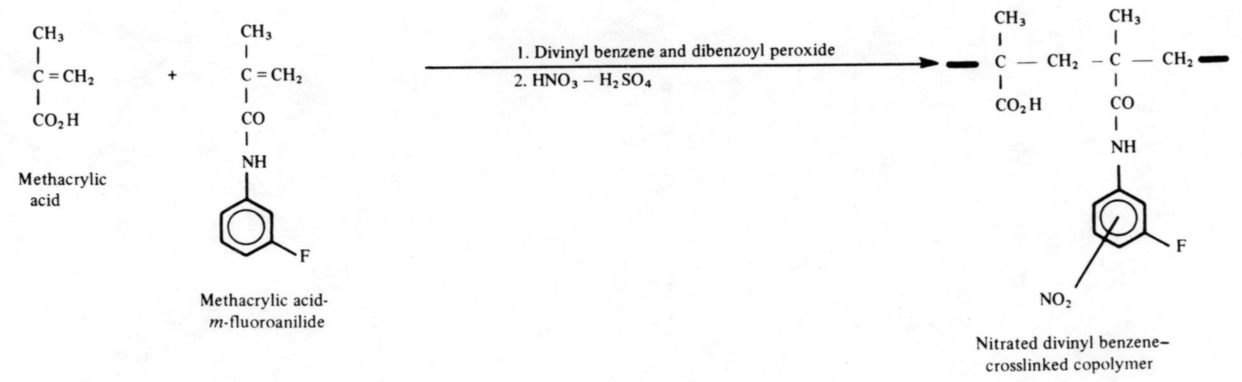

Methacrylic acid + Methacrylic acid-*m*-fluoroanilide

1. Divinyl benzene and dibenzoyl peroxide
2. $HNO_3 - H_2SO_4$

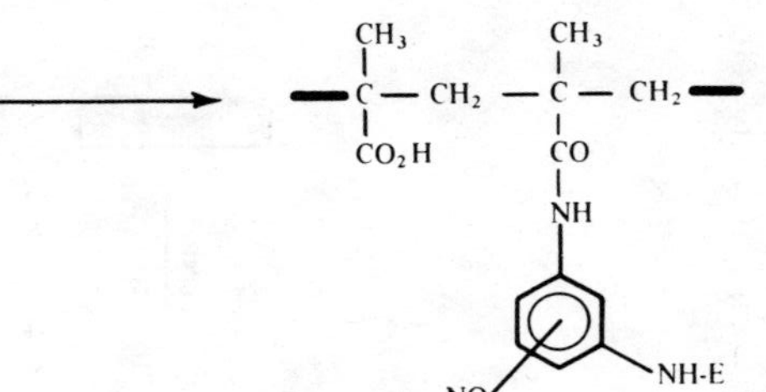


Nitrated divinyl benzene–crosslinked copolymer

(20)

**Coupling:**

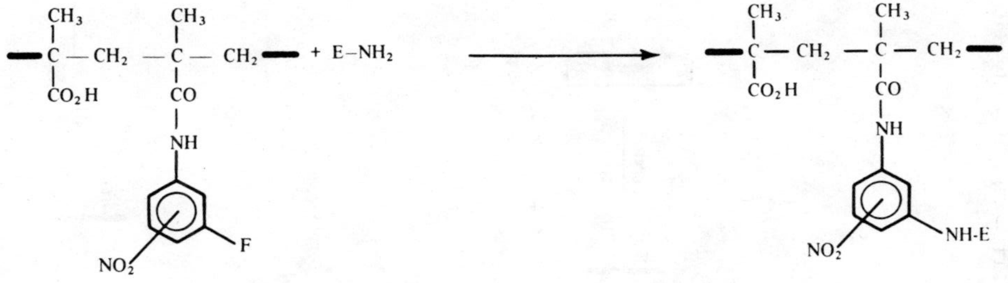

$-C(CH_3)(CO_2H)-CH_2-C(CH_3)(CONH-C_6H_3(NO_2)F)-CH_2- \; + \; E-NH_2 \longrightarrow -C(CH_3)(CO_2H)-CH_2-C(CH_3)(CONH-C_6H_3(NO_2)(NH\text{-}E))-CH_2-$ (21)

$$-CH_2-CH_2-CH-CH-CH_2-CH_2-CH-CH-CH_2-CH_2-CH-CH- \text{ (anhydride rings) } + E-NH_2 + NH_2-(CH_2)_6-NH_2 \longrightarrow$$

$$\text{Crosslinked product: } -CH_2-CH_2-CH(C{=}O)-CH(CO_2^-)-CH_2-CH_2-CH(C{=}O)-CH(CO_2^-)-CH_2-CH_2-CH(CO_2^-)-CH(CO_2^-)-$$

with C=O–NH–$(CH_2)_6$–NH–C=O, C=O–NH–E–NH–C=O and NH-E–C=O linkages to a second chain $-CH_2-CH_2-CH(C{=}O)-CH(CO_2^-)-CH_2-CH_2-CH(C{=}O)-CH(CO_2^-)-CH_2-CH_2-CH(C{=}O)-CH(CO_2^-)-$ (22)

Activation:

$$\text{matrix}\begin{matrix}-OH\\-OH\end{matrix} \xrightarrow[-HBr]{CNBr} \left[\text{matrix}\begin{matrix}-O-C\equiv N\\-OH\end{matrix}\right] \begin{matrix}\xrightarrow{H_2O} \text{matrix}\begin{matrix}-O-\overset{O}{\overset{\|}{C}}-NH_2\\-OH\end{matrix} \text{ Non-reactive carbamate}\\ \uparrow H_2O \\ \longrightarrow \text{matrix}\begin{matrix}-O\\-O\end{matrix}\!\!>C=NH \text{ Reactive imidocarbonate}\end{matrix} \quad (23)$$

Coupling:

$$\text{matrix}\begin{matrix}-O\\-O\end{matrix}\!\!>C=NH \xrightarrow{E-NH_2} \begin{cases}\text{matrix}\begin{matrix}-O-\overset{NH}{\overset{\|}{C}}-NH-E\\-OH\end{matrix} & \text{Isourea}\\ \text{matrix}\begin{matrix}-O\\-O\end{matrix}\!\!>C=N-E & N\text{-Substituted imidocarbonate}\\ \text{matrix}\begin{matrix}-O-\overset{O}{\overset{\|}{C}}-NH-E\\-OH\end{matrix} & N\text{-Substituted carbamate}\end{cases} \quad (24)$$

registered trademark for spherical *dextran* gel particles produced by Pharmacia Fine Chemicals, Inc. Sephadex is a modified dextran whose molecules are crosslinked to produce networks of varying degrees. Other manufacturers also produce agarose gels. The immobilization method is simple and reliable, and the attachment of the enzyme to the matrix is performed under mild conditions. The method involves the activation of the polysaccharide with cyanogen bromide (chloride or iodide) to give the reactive imidocarbonate which subsequently reacts with the protein. The chemistry of the activation and coupling steps is given in Equations 23 and 24. The chemistry of

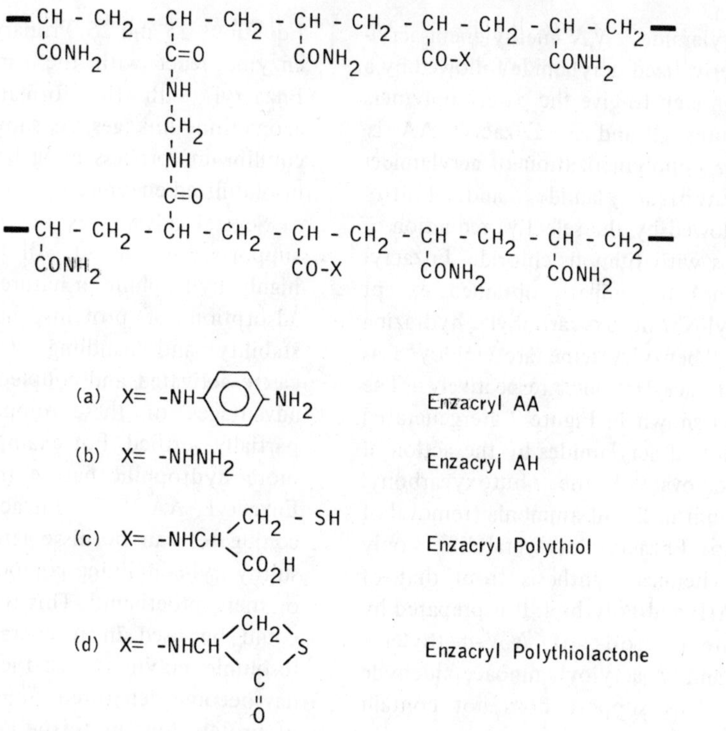

$$-CH-CH_2-CH(C{=}O-NH-CH_2-NH-C{=}O-)-CH_2-CH(CONH_2)-CH_2-CH(CO\text{-}X)-CH_2-CH(CONH_2)-CH_2-$$

(a) X= -NH-C₆H₄-NH₂ — Enzacryl AA

(b) X= $-NHNH_2$ — Enzacryl AH

(c) X= $-NHCH(CH_2-SH)(CO_2H)$ — Enzacryl Polythiol

(d) X= -NHCH< (cyclic thiolactone: $CH_2$, S, C=O) — Enzacryl Polythiolactone

FIGURE 2. Structure of Enzacryl AA, AH, Polythiol, and Polythiolactone.

the individual steps shown in these equations is based mainly on observations made with appropriate model compounds.[48] The method was first described by Axen, Porath, and Ernback in 1967 for the attachment of α-chymotrypsin to Sephadex[71] and agarose.[72]

### *Newer Methods for the Covalent Attachment of Enzymes to Water-Insoluble Functionalized Supports*

This section describes several recently reported water-insoluble supports or activation procedures for previously described matrices. The methods discussed are only representative examples of many that have been described in the literature and which are given in Table 4 at the end of this chapter. These were selected because they appear to be applicable to the insolubilization of enzymes through various amino acid residues, the procedures are simple, the chemistry is somewhat unusual, and the results obtained so far are quite encouraging.

### *Enzacryl® Supports*

A recently introduced series of water-insoluble supports for enzyme immobilization are the Enzacryl® type of matrices.[56,73–76] The supports are copolymers of acrylamide and various derivatives of acrylamide. They are commercially available under the Koch-Light Laboratories, Ltd. trademark of Enzacryl and are further designated as Enzacryl AA, AH, Polythiol, Polythiolactone or Polyacetal. Their basic structures are shown in Figures 2 and 3.

The chemical synthesis of Enzacryl supports is straightforward and consists of initial copolymeri-

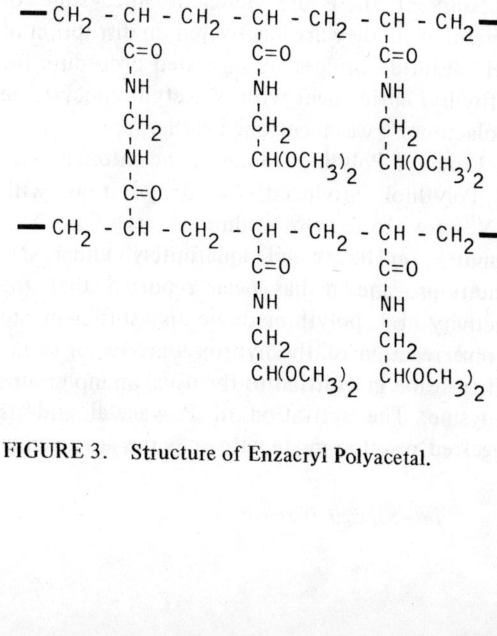

$$-CH_2-CH(C{=}O-NH-CH_2-NH-C{=}O-)-CH_2-CH(C{=}O-NH-CH_2-CH(OCH_3)_2)-CH_2-CH(C{=}O-NH-CH_2-CH(OCH_3)_2)-CH_2-$$

$$-CH_2-CH-CH_2-CH(C{=}O-NH-CH_2-CH(OCH_3)_2)-CH_2-CH(C{=}O-NH-CH_2-CH(OCH_3)_2)-CH_2-$$

FIGURE 3. Structure of Enzacryl Polyacetal.

zation of acrylamide, $N,N'$-methylenebisacrylamide, and a derivatized acrylamide followed by a simple activation step to give the parent polymers shown in Figures 2 and 3. Enzacryl AA is prepared by the copolymerization of acrylamide, $N,N'$-methylenebisacrylamide, and 4-nitroacrylanilide followed by the selective reduction of the nitro groups with titanous chloride. Enzacryl AH and Polythiol are similarly obtained, except that $N$-acryloyl-$N'$-$t$-butoxycarbonyl hydrazine and $N$-acryloyl-$S$-benzylcysteine are employed as the derivatized acrylamides, respectively. The parent polymers shown in Figure 2 are generated from the substituted acrylamides by the action of dilute acid (removal of the $t$-butoxycarbonyl group) and sodium in liquid ammonia (removal of the benzyl group). Enzacryl Polyacetal differs only slightly in its chemical synthesis from that of Enzacryl AA, AH, and Polythiol. It is prepared by the copolymerization of $N,N'$-methylenebisacrylamide and $N$-acryloylaminoacetaldehyde dimethylacetal. This support does not contain acrylamide.[56,76]

The chemical transformation of Enzacryl AA and AH into active polymers which combine with amino acid residues of enzymes is the same as for other supports discussed previously. Enzacryl AA has been converted to both the diazonium and isothiocyanato derivatives. Enzacryl AH, on the other hand, can only be converted to one active enzyme coupling group – the acid azide function. Covalent bonding of enzymes with Enzacryl Polythiol is accomplished readily, and often spontaneously, by oxidative coupling of a sulfhydryl-containing protein with the free sulfhydryl groups of the polymer. Prior sulfhydryl enrichment of the enzyme or protein may be necessary if these are deficient in cystine or cysteine or if they are inactivated on disruption of their disulfide bridges. A suggested procedure for sulfhydryl enrichment with $N$-acetylhomocysteine thiolactone[56] was mentioned earlier (see Equation 3). Enzacryl Polythiolactone is the internal ester of Polythiol produced by dehydration with $N,N'$-dicyclohexylcarbodiimide (DCC). This support can be stored indefinitely under dry conditions, and it has been reported that the reactivity of polythiolactone is sufficient to involve reaction of the hydroxyl groups of serine and tyrosine in addition to the usual amino groups of lysine. The activation of Polyacetal and its suggested reaction mode with enzymes are given in Equations 25 and 26. Primary amino groups of the enzyme react with the carbonyl groups of the Enzacryl with the formation of aminol and azomethine linkages. As shown, the reaction is an equilibrium process lying heavily in favor of the insolubilized enzyme.

Several advantages of these various Enzacryl supports are claimed. All Enzacryl supports are highly hydrophilic in nature, exhibit low physical adsorption of proteins, have good mechanical stability and handling characteristics, and are easily activated and coupled with enzymes. Some advantages of these supports have also been partially verified. For example, Enzacryl AH has a more hydrophilic nature than phenyl-containing Enzacryl AA.[56,75] Enzacryl Polythiol-enzyme conjugates can be dissociated by treatment with sulfhydryl-containing compounds such as cysteine or mercaptoethanol. This reversibility of coupling could be used in regenerating an active water-insoluble polymer once the immobilized enzyme has become denatured. Similarly, the reversibility of protein binding to the Polyacetal carrier could be highly advantageous. With this support, the dissociation of the enzyme conjugate could be accomplished, in principle at least, by treatment with dilute acids or suitable buffers. To date, however, these support-regeneration methods have proved unsuccessful. Dissociation of enzymes from the support can be accomplished with suitable *macromolecular* substrates of the immobilized enzymes. For example, partial dissociation of the trypsin derivative was achieved by repeated washing of the enzyme-polymer conjugate with casein solution. Similarly, immobilized $\alpha$-amylase and dextranase were partially dissociated by repeated washings with starch and dextran solutions, respectively.[56]

Protein coupling ranges from about 2 to 50 mg of enzyme/g of carrier.[73,76] Surprisingly, the amount of enzyme bound to Enzacryl Polyacetal depends more on the water regainability of the polymer than on its aldehydic content.[56,76]

### *Modified Polyacrylamide Gel Beads*

Two publications appeared recently describing the chemical transformation of preformed polyacrylamide gel beads into reactive supports capable of covalent bond formation with enzymes. Polyacrylamide gel beads have a crosslinked structure composed of acrylamide and $N,N'$-methylenebisacrylamide, and as a support,

$$\text{—CH–CH}_2\text{—} \; | \; \text{CO} \; | \; \text{NH} \; | \; \text{CH}_2 \; | \; \text{CH(OCH}_3)_2 \xrightarrow{H^+} \text{—CH–CH}_2\text{—} \; | \; \text{CO} \; | \; \text{NH} \; | \; \text{CH}_2 \; | \; \text{CHO} + CH_3OH \qquad (25)$$

$$\text{—CH–CH}_2\text{—} \; | \; \text{CO} \; | \; \text{NH} \; | \; \text{CH}_2 \; | \; \text{CHO} + E-NH_2 \rightleftharpoons \text{—CH–CH}_2\text{—} \; | \; \text{CO} \; | \; \text{NH} \; | \; \text{CH}_2 \; | \; \text{CH(OH)–NH–E} \rightleftharpoons \text{—CH–CH}_2\text{—} \; | \; \text{CO} \; | \; \text{NH} \; | \; \text{CH}_2 \; | \; \text{CH=N–E} \qquad (26)$$

offer the advantage of good chemical stability, a uniform physical state and porosity, and commercial availability. Polyacrylamide is also a neutrally charged carrier. Although polyacrylamide is chemically stable, resistant to hydrolysis in the pH range between 1 and 10, and does not react with nitrous acid (as do primary amides), it can be chemically transformed by several reactants. The chemical transformations are simple to conduct and subsequent coupling of an enzyme is likewise easily achieved. In 1969, Inman and Dintzis[77] described a series of modification reactions of preformed polyacrylamide gel beads with diamines such as hydrazine and ethylenediamine. A large excess of the amine is used, and the reaction is an amine exchange (Equations 27 and 28).

$$\text{Polyacrylamide}\; | -CONH_2 + H_2NCH_2CH_2NH_2 \longrightarrow | -CONHCH_2CH_2NH_2 + NH_3 \qquad (27)$$

$$| -CONH_2 + H_2NNH_2 \longrightarrow | -CONHNH_2 + NH_3 \qquad (28)$$

A temperature of approximately 50° is required for hydrazide formation. Although not shown, these amino-containing polyacrylamides were further derivatized to a host of reactive supports that could be used for enzyme immobilization, affinity chromatography, and ion-exchange separations. The formed hydrazide shown in Equation 28 was treated with nitrous acid to give the acid azide. Trypsin, immobilized on this support, retained considerable activity; protein binding was estimated to be about 170 mg/g of support. No other examples of the covalent attachment of enzymes were given. Both the aminoethyl and hydrazide polyacrylamide gel beads are now commercially available (Bio-Rad, Aminoethyl and Hydrazide Bio-Gel® P) and should prove to be excellent supports for enzyme insolubilization.

Another modification of preformed polyacrylamide gel beads, based on the reaction with glutaraldehyde, was described by Weston and Avrameas.[78] Although the chemistry of the method has not been established, it has been suggested that glutaraldehyde, when used in excess, reacts monofunctionally with free amino or amide groups present in polyacrylamide to give a modified support that contains aldehydic groups available for coupling with enzymes. Activation of the polyacrylamide with the dialdehyde consists of incubating the mixture at 33° for 17 hr at a pH of 6.9 or lower. The beads are washed and the enzyme is added to achieve coupling. Acid phosphatase, ribonuclease, glucose oxidase, trypsin, and chymotrypsin were immobilized in this manner. Relatively good activity and protein binding were observed.

***Poliodal and Copoliodal Supports***

In a series of papers, Brown, Racois, and

colleagues described new synthetically produced supports for the covalent attachment of enzymes.[43,79-82] The supports are homo- and copolymers of either 4-iodobutyl methacrylate, 2-iodoethyl methacrylate and methyl methacrylate and have been named "Poliodals." Poliodal-4 is the homopolymer of 4-iodobutyl methacrylate, Poliodal-2 is the homopolymer of 2-iodoethyl methacrylate, and variously designated Copoliodals are the copolymers of either the 4-iodobutyl or 2-iodoethyl derivatives and methyl methacrylate. The preparation and properties of these new enzyme insolubilization supports were also disclosed.[81] Synthesis of the reactive alkyl iodide is achieved through the direct polymerization of the iodoalkyl monomer or through a halide exchange reaction of the appropriate chloride with sodium iodide in acetone. The reactions for the preparation of Poliodal-4 are given in Equation 29 . Covalent coupling of enzymes is achieved in slightly alkaline buffered solutions (pH 8-9) and seems to occur via an alkylation of predominantly the $\epsilon$-amino group of lysyl residues with the polymeric alkyl iodides (Equation 30). Other nucleophilic groups of enzymes could also participate in covalent bond formation. Relatively good protein binding is obtained with these supports. With $\alpha$-chymotrypsin, maximum binding occurred in 24 hr at 2° and pH 8 with binding decreasing in the order of Poliodal-4, Poliodal-2, and Copoliodals, respectively.[80]

### *Dialdehyde Starch-Methylenedianiline Resin (S-MDA)*

A new type of highly insoluble diazotizable support, dialdehyde starch-methylenedianiline (S-MDA), was described by Goldstein et al.[47] in 1970. The resin is prepared by the condensation of dialdehyde starch, a commercially available periodate-oxidation product of starch, with 4,4′-diaminodiphenylmethane (bismethylene-dianiline) followed by reduction of the resulting Schiff base with sodium borohydride (see Equation 31). Activation is achieved with nitrous acid to give the corresponding diazonium derivative. Covalent coupling of several enzymes (papain, subtilopeptidase A, and polytyrosyl trypsin) occurred readily through covalent bond formation with tyrosine, arginine, and lysine residues. Protein binding was substantial with all enzymes and reached a level of 290 mg of protein/g of resin for polytyrosyl trypsin. The water-insoluble S-MDA protein conjugates are of particulate form, are easily filtered, and can be used in columns. The use of this crosslinked, versatile support should increase in the future.

$$CH_2=C(CH_3)-CO_2-(CH_2)_4-Cl \xrightarrow[\text{Acetone}]{\text{NaI}} CH_2=C(CH_3)-CO_2-(CH_2)_4-I$$

Benzoyl peroxide or Azobisisobutyronitrile (polymerization, both monomers)

$$-CH_2-C(CH_3)(CO_2(CH_2)_4-Cl)- \xrightarrow[\text{Acetone}]{\text{NaI}} -CH_2-C(CH_3)(CO_2(CH_2)_4-I)- \quad (29)$$

Poliodal-4

$$-CH_2-C(CH_3)(CO_2(CH_2)_3-CH_2-I)- + E-NH_2 \longrightarrow -CH_2-C(CH_3)(CO_2(CH_2)_3-CH_2-NH-E)- \quad (30)$$

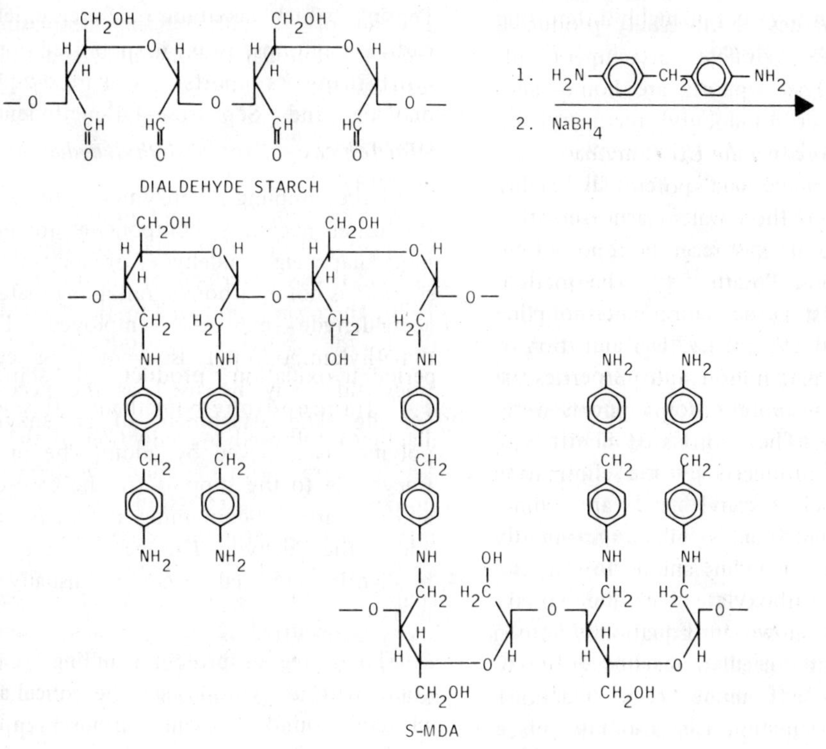

DIALDEHYDE STARCH

1. $H_2N-C_6H_4-CH_2-C_6H_4-NH_2$

2. $NaBH_4$

S-MDA

(31)

### *Transition Metal-Activated Supports*

A very recent development in water-insoluble supports for the immobilization of enzymes was disclosed by Barker, Emery, and Novais[83] in 1971. The method is based on the activation of natural or synthetic matrices such as cellulose, nylon, or glass with salts of transition metals. The most commonly used transition metal salt is $TiCl_4$ but chlorides of titanium, tin, zirconium, vanadium, and iron can be employed. The activation procedure consists of steeping the support in a solution of the transition metal salt. The choice of the particular salt and its concentration are dependent on the support. Steeping is followed by filtration of the activated matrix and washing it with buffers to remove any excess salt. Enzyme coupling is achieved by contacting the protein with the support usually in a buffer solution at the optimum pH of the enzyme. The solid-metal-enzyme conjugate is then washed to remove any adsorbed enzyme. Although titanium salts are generally the most reactive, some supports are degraded by the low pH of solutions of titanous and titanic chloride. Relative enzymatic activities of the bound enzymes and protein binding are relatively high. For example, glucoamylase bound on microcrystalline cellulose through various transition metal salts exhibited a specific activity of 45 to 74% with a protein binding of 8.1 to 147 mg/g of enzyme-polymer conjugate. The method has been used for the binding of several enzymes to cellulose, nylon, glass, and even to yeast cells.[83]

The exact nature of the chemical interaction or type of bonds between the support and metal and the metal and enzyme have not yet been elucidated. Almost certainly, the metal-support and metal-protein bonds are at least partially covalent in nature. Results from studies of the reaction of both model diols and cellulose and ethylcellulose with $TiCl_4$ suggest possible chemical reactions which could be responsible for the activation of the cellulose and the coupling of the enzyme to the transition metal complex[84] (see Equations 32 and 33). The reactions shown are the simplest ones that can be envisioned to occur with cellulose and enzymes. The activation step could also involve further reactions of the titanium complex with additional hydroxyl groups or with other oxygens of the polysaccharide. Likewise, numerous functional groups of an enzyme could participate in bond formation with the activated support.

The method appears to be reversible. The full

potential and versatility of this highly interesting method remain to be explored.

### *Activation of Supports via the Ugi Reaction*

An extremely novel and potentially highly versatile approach to the covalent attachment of enzymes to polymers was recently reported by Axen, Vretblad, and Porath.[85,86] The method involves an interesting activation and coupling procedure that is based on the "Ugi" reaction or rearrangement.[87] Equations 34 to 36 give the proposed scheme for the steps involved in product formation. The protonated Schiff base (immonium ion), produced by the elimination of water from a carbonyl- and an amino-containing compound, adds in an apparently concerted fashion with an anion to an isocyanide. With a carboxylate ion, the ternary addition complex shown in Equation 34 then undergoes an intramolecular rearrangement to give the substituted amide (Ugi reaction). Hydroxyl ion participation can also take place (Equation 36). The overall reaction for the formation of the amide involves four different functionalities – carbonyl, amino, isocyano, and carboxyl.

The possibility of success of forming a product in high yield in a chemical transformation involving at least four different and separate organic functionalities could be questioned. However, it is precisely this large number of functionalities that gives this method the numerous variations that can be employed. The substances to be attached to the carrier polymer may contain amino, carbonyl, or isocyano groups; with enzymes, the functional groups that are utilized for covalent bond formation are only the amino and carboxyl groups. The reaction occurs in aqueous solution and is mild enough to be used for enzyme insolubilizations. A number of polymers have been employed for the immobilization of pepsin[86] and $\alpha$-chymotrypsin.[85] For example, $\alpha$-chymotrypsin was immobilized on carboxyl-containing (Sephadex and agarose), amino-containing (lysyl-Sepharose, AE-polyacrylamide, and Enzacryl AA), carbonyl-containing (various oxidized Sepharoses), and amino- and carboxyl-containing (keratin and wool) supports. In fact, $\alpha$-chymotrypsin was even attached to previously immobilized papain.[85] Both enzymes retained their activity toward their normal substrates. Pepsin, which contains a preponderance of carboxyl groups, was immobilized on amino-containing supports (Sepharose-*p*-phenylenediamine and Sepharose-4,4′-methylenedianiline conjugates).

In the coupling of enzymes to these supports, the other necessary components are either low molecular weight aldehydes or isocyanides. If the support is not a carbonyl-containing material, then acetaldehyde may be employed. Either 3-dimethylaminopropyl isocyanide or cyclohexyl isocyanide may be used as the necessary isocyanide. The attachment of an enzyme to a polymer is achieved by adding the protein and isocyanide to the support (in the case of amino- and/or carboxyl-containing polymers) and then adding the aldehyde. The reactions are conducted in slightly acid medium and are usually complete in 6 hr.

The degree of protein coupling to carriers is good. With $\alpha$-chymotrypsin, the typical amount of protein bound to the various supports was approximately 40 to 80 mg/g of enzyme-polymer conjugate. Values as high as 395 and as low as 2 mg/g were also reported. Activities of the insolubilized $\alpha$-chymotrypsin were greatest with the carbonyl-containing supports.

### **Copolymerization of Enzymes with Reactive, Low Molecular Weight Monomers**

The insolubilization of enzymes through the copolymerization of proteins with low molecular weight reactive monomers has been described only a few times in the literature, and all examples cited have involved the copolymerization of the enzyme with reactive *N*-carboxyamino acid anhydrides. The preparation of polypeptidyl proteins was first described by Stahmann and Becker,[88] who showed that the polymerization of the reactive anhydrides may be initiated by the amino groups of proteins to yield the corresponding polypeptidyl derivatives. Peptide polymerization involves the $\alpha$- and $\epsilon$-amino groups of enzymes and consists of a chain elongation on those residues (Equation 37). The polypeptidyl derivatives originally described by Stahmann and Becker[88] and subsequently by Katchalski[52,53] and coworkers were only slightly modified proteins that still were water-soluble. A water-soluble polytyrosyl trypsin derivative was mentioned previously on page 9. Yet, under appropriate condi-

**Activation:**

$$\text{Cellulose-OH} + TiCl_4 \longrightarrow \text{Cellulose-O-}TiCl_3 + HCl \qquad (32)$$

**Coupling:**

$$\text{Cellulose-O-}TiCl_3 + \text{E-OH} \longrightarrow \text{Cellulose-O-Ti(Cl)}_2\text{-O-E} \qquad (33)$$

$$\begin{matrix} R_1 \\ R_2 \end{matrix}\!\!>C=O + H_2N-R_3 \xrightarrow{H^+} \begin{matrix} R_1 \\ R_2 \end{matrix}\!\!>C=\overset{+}{N}H-R_3 + H_2O \qquad (34)$$

$$R_4-N\equiv C + {}^{+}C(R_1)(R_2)-NH-R_3 + {}^{-}O-\overset{O}{\overset{\|}{C}}-R_5 \longrightarrow R_4-N=C(-C(R_1)(R_2)-NH-R_3)(-O-\overset{O}{\overset{\|}{C}}-R_5) \xrightarrow{\text{Rearrangement}} R_4-NH-\overset{O}{\overset{\|}{C}}-C(R_1)(R_2)-N(R_3)-\overset{O}{\overset{\|}{C}}-R_5 \qquad (35)$$

Ugi Reaction

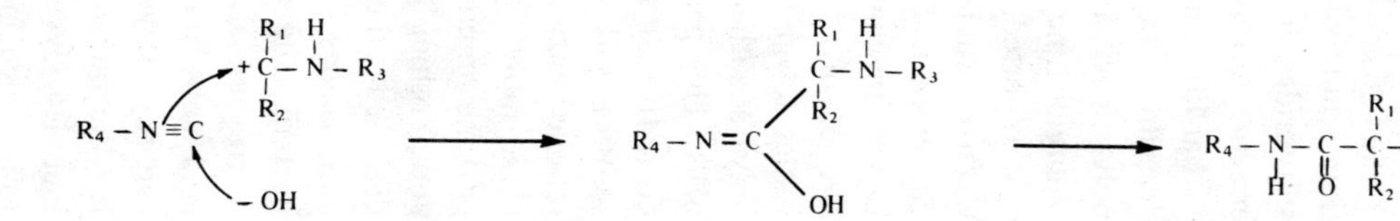

$$\begin{array}{c}NH_2\\|\\E\\ \diagup\ \diagdown\\ NH_2\quad NH_2\end{array} + \begin{array}{c}R\\|\\CH\\HN\quad C=O\\ \quad\ O\\ C\\ \|\\ O\end{array} \longrightarrow \begin{array}{c}A\\|\\NH\\|\\E\\ \diagup\ \diagdown\\ A-HN\quad NH-A\end{array} \quad (37)$$

$$\text{where } A = -\left(\underset{\underset{O}{\|}}{C}-\overset{\overset{R}{|}}{CH}-NH\right)_n \underset{\underset{O}{\|}}{C}-\overset{\overset{R}{|}}{CH}-NH_2$$

tions and with suitable *N*-carboxyamino acid anhydrides, it is possible to obtain water-insoluble enzyme derivatives.

Bar-Eli and Katchalski,[52] in addition to describing the preparation of a water-insoluble polytyrosyl trypsin – diazotized *p*-amino-DL-phenylalanine and L-leucine enzyme conjugate, also reported, with Glazer, the preparation of a water-insoluble trypsin derivative prepared by the copolymerization of the enzyme with *N*-carboxy-L-tyrosine anhydride.[53] This polytyrosyl derivative dissolved readily in aqueous solution of pH values lower than 5 or higher than 9.5, but had only a rather limited solubility between these pH's. This partially insoluble polytyrosyl trypsin derivative, insolubilized by precipitation from solution with acetate, phosphate, or borate anions, retained enzymic activity upon redissolution.

A more thorough study of this method of enzyme immobilization was reported by Kirimura and Yoshida[89] in 1966. In their patent, they disclosed the preparation of water-insoluble aminoacylases (derived from such sources as fungi, bacteria, and animal tissues) through the copolymerization of the enzymes with *N*-carboxyamino acid anhydrides. Typical amino acids whose *N*-carboxy anhydride derivatives could be employed for the preparation of water-insoluble derivatives included glycine, alanine, $\alpha$-amino-*n*-butyric acid, valine, leucine, isoleucine, phenylalanine, and several $\beta$- and $\gamma$-esters of aspartic and glutamic acid, respectively. However, *N*-carboxy anhydrides of proline, serine, and threonine could not be used because the resulting derivatives were still soluble. With regard to the solubility or insolubility of these modified polypeptidyl acylases prepared by these investigators, it should be emphasized that they often referred to their supposed insoluble enzyme derivatives as "practically insoluble in water." The reported insolubility of these immobilized enzymes should be viewed with caution.

The immobilized acylases retained their activity and exhibited no loss in activity upon storage in buffer solutions at room temperature for two weeks. The water-insoluble enzyme derivatives could be lyophilized. The method offers an interesting approach for producing water-insoluble enzyme conjugates.

### Water-Soluble Enzyme-Polymer Conjugates

Two reports have appeared recently describing the preparation of covalently linked, water-*soluble,* enzyme-polymer conjugates. These water-soluble enzyme derivatives were purposely prepared in order (1) to increase the effective molecular size of the parent enzyme so that it could be used in an ultrafiltration cell without problems of enzyme leakage, and (2) to chemically modify the enzyme so that it would have superior physical properties. Wykes, Dunnill, and Lilly[90] prepared water-soluble, enzyme-polymer derivatives of $\alpha$-amylase from soluble dextran, DEAE-dextran, and CM-cellulose. The water-soluble CM-cellulose derivative did not pass through the membrane of the ultrafiltration cell and showed enhanced thermal stability. Likewise, these investigators along with O'Neill[91] described a water-soluble $\alpha$-chymotrypsin-dextran conjugate. It should be mentioned that the covalent attachment of an enzyme to a *water-soluble* support does not constitute a method for immobilizing enzymes; as will be seen later, the immobilization of such enzymes is accomplished only with the membrane-dependent device. The use of ultrafiltration cells as a means of immobilizing enzymes is discussed in Chapter 8.

Several additional water-soluble, enzyme-polymer conjugates have been described briefly. These were often not prepared purposely for eventual use in membrane-dependent devices and were not the desired aim or result of the investigators. For example, water-soluble enzyme polymer derivatives were obtained by the reaction of the protein with maleic anhydride-ethylene copolymers[68] and polymethacrylic acid anhydride[92] when no crosslinking reagent (1,6-diaminohexane) was employed; effective crosslinking is essential for producing water-insoluble enzyme-polymer conjugates. Likewise, attempted insolubilization of apyrase with polyvinylamine[93] via glutaric acid dihydrazide gave a water-soluble derivative. Some other water-soluble enzyme-polymer conjugates described are α-chymotrypsin on CM-cellulose via the acylazide derivative[57] and α-chymotrypsin on polyacrylic acid,[94] CM-cellulose[94] and poly-L-glutamic acid[94] via coupling with *N*-ethyl-5-phenylisoxazolium-3′-sulfonate (Woodward's Reagent K). The activation and coupling reactions are given in Equations 38 and 39.

Water-*soluble* α-chymotrypsin derivatives have also been prepared from water-*insoluble* enzyme-polymer conjugates (Sephadex G-200 bound) through the degradative action of dextranase.[95] Dextranase catalyzes the hydrolysis of the macromolecular dextrans to give water-soluble oligosaccharides of various lengths, and Sephadex G-100, 150, and 200 can be completely solubilized by the enzyme.

**Physical Nature of Water-Insoluble Matrices**

The physical forms of the numerous supports used for the immobilization of enzymes are rather diverse. Most have a form that allows easy filtration or centrifugation, and they have a large surface area to volume or weight ratio. The latter is desirable in order to achieve a high degree of chemical modification of the support and a corresponding high protein to carrier ratio. Most often, the supports are powdery, granular, spongy, gelatinous, or fibrous in nature, and many possess a spherical bead-like structure.

Several publications have also described

Activation:

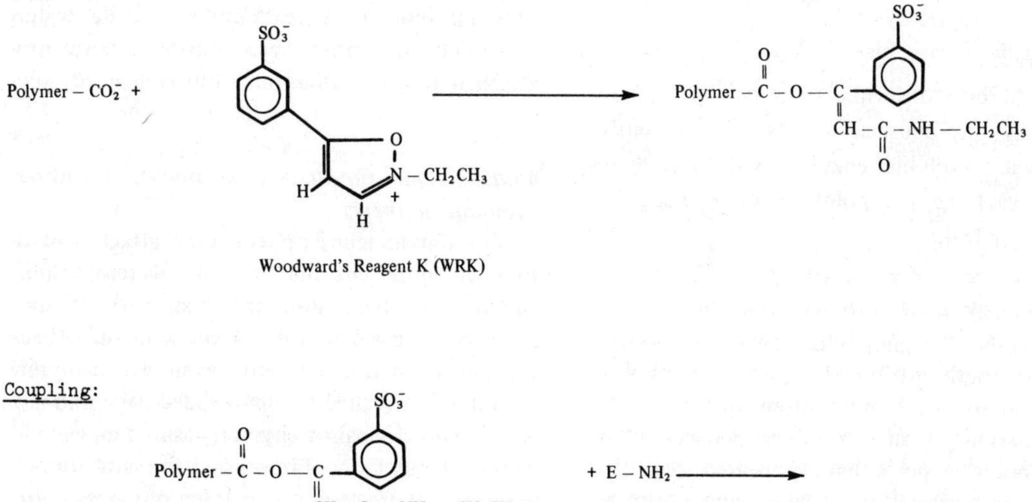

Polymer – $CO_2^-$ + [Woodward's Reagent K: $SO_3^-$-phenyl isoxazolium, N – $CH_2CH_3$] ⟶ Polymer – C(=O) – O — C(aryl-$SO_3^-$)=CH – C(=O) – NH — $CH_2CH_3$ (38)

Woodward's Reagent K (WRK)

Coupling:

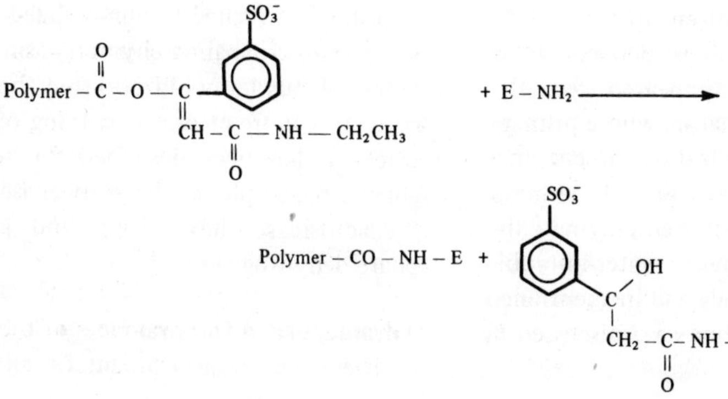

Polymer – C(=O) – O — C(aryl-$SO_3^-$)=CH – C(=O) — NH — $CH_2CH_3$ + E – $NH_2$ ⟶ (39)

Polymer – CO – NH – E + (aryl-$SO_3^-$)C(OH)=… $CH_2$ – C(=O) – NH — $CH_2CH_3$

supports of somewhat unusual character. Enzymes were covalently attached to such cellulosic materials as paper sheets of various types (plain filter paper, Whatman No. 1,[40,62,64] and anion-exchange paper, Whatman DE-81[40,62-64,96,97]), dialysis tubing,[98] cellophane membranes,[99] and cheesecloth.[100] Several enzymes (trypsin and β-fructofuranosidase) were attached to the inside surfaces of nylon and polyaminostyrene tubes. In the case of trypsin,[101] the inside surface of a nylon tube (4.0 cm long, 0.1 cm i. d.) was partially hydrolyzed with acid, treated with nitrous acid to destroy the newly created amino groups, and then the remaining carboxyl groups were modified with either benzidine or hydrazine in the presence of DCC to give the aminoaryl or hydrazide derivative, respectively. Activation of both derivatives was accomplished with nitrous acid. In the case of β-fructofuranosidase,[102] polystyrene tubes (0.2 cm i. d. ) were nitrated, then reduced to the corresponding amino derivatives and subsequently diazotized to give the active support. L-Asparaginase was attached to a Dacron® vascular prosthesis,[103] glucose oxidase was coupled to nickel oxide screens,[104] and glutamate dehydrogenase was bonded to thin films of collagen.[105]

**Miscellany**

It might be worthwhile to point out certain noteworthy or unusual reports of covalently linked water-insoluble enzymes with regard to their preparation. The points considered are included in Table 4.

*Combination Methods of Immobilization*

Frequently, the immobilization of an enzyme cannot be strictly attributed to purely chemical or physical means, and quite often the successful immobilization of an enzyme is dependent on both. Certain reports have appeared recently, however, from several investigators whose primary aim was to achieve immobilization of enzymes through the simultaneous use of several methods. Only combination methods employing the covalent attachment of enzymes to water-insoluble polymers as one of the methods will be mentioned here; other combination methods are discussed in later chapters.

A combination covalent attachment-physical entrapment of enzymes was described by Mosbach[106] and Mosbach and Mattiasson[107] in 1970. The enzymes, trypsin or glucose-6-phosphate dehydrogenase, were immobilized by entrapment within a forming polyacrylamide gel and by the covalent attachment through a water-soluble carbodiimide. Specifically, the procedure for preparing immobilized trypsin went as follows: the enzyme and *N,N'*-methylenebisacrylamide were dissolved in phosphate buffer; after addition of acrylamide and acrylic acid, 1-cyclohexyl-3-(2-morpholinoethyl)-carbodiimide metho-*p*-toluenesulfonate (CMC) was added; finally, the polymerization catalyst (ammonium persulfate and β-dimethylaminopropionitrile) was added, and the polymerization proceeded to give the desired immobilized enzyme derivative. Immobilized glucose-6-phosphate dehydrogenase was similarly prepared except that the polymerization catalyst consisted of ammonium persulfate and *N,N,N',N'*-tetramethylethylenediamine (TEMED) and that the carbodiimide was the last reagent added. Equation 40 shows the covalent coupling reaction.

Another covalent combination method, reported by Marshall and Falb,[108] consisted of the physical adsorption of the enzyme onto collodial silica particles followed by its intermolecular crosslinking with a bifunctional reagent to give a water-insoluble enzyme "envelope." The water-insoluble derivative was further covalently attached to a water-insoluble functionalized polymer.

*Multi-Enzyme Covalent Attachments to Water-Insoluble Supports*

The simultaneous or sequential attachment of two or more enzymes to one water-insoluble support has been disclosed. Examples of such systems are given in Table 4 and some have been previously mentioned. For example, α-chymotrypsin was attached to immobilized papain to give a Sepharose-papain-α-chymotrypsin conjugate.[85] Simultaneous immobilization on the same support of two or more enzymes carrying out *consecutive* reactions has been described for hexokinase and glucose-6-phosphate dehydrogenase[107] and for β-galactosidase, hexokinase, and glucose-6-phosphate dehydrogenase.[109]

**Advantages and Disadvantages of the Method**

The covalent attachment of an enzyme to a

$$\text{Polymer–CO}_2\text{H} + \text{R–N=C=N–R} \longrightarrow \text{Polymer–}\overset{\overset{\text{O}}{\|}}{\text{C}}\text{–O–}\underset{\underset{\text{N–R}}{\|}}{\overset{\overset{\text{NHR}}{|}}{\text{C}}} \xrightarrow{\text{E–NH}_2} \text{Polymer–}\overset{\overset{\text{O}}{\|}}{\text{C}}\text{–NH–E} + \text{RNHCONHR} \qquad (40)$$

Carbodiimide *O*-Acylisourea

water-insoluble support offers numerous advantages as a method for immobilizing enzymes. The coupling of the enzyme to a functionalized support is experimentally easy to conduct and consists of adding the enzyme to a suspension of the polymer, allowing the coupling to proceed, and removing any noncovalently bound enzyme by suitable washings. The solid-supported catalysts are easy to handle in manipulative operations; they can be filtered or centrifuged and resuspended. Such easy removal of the catalyst allows quick termination of a reaction at a particular stage and permits selective or partial transformations to be achieved. Further, the product is not contaminated with the enzyme. In addition, because of the variations in the chemical and physical nature of the support, these immobilized enzymes can be adapted to a variety of specific engineering requirements. They can be used in a continuous fashion in packed-bed, stirred-tank, or fluidized-bed reactors. Various physical forms of the support material can also be employed; the enzyme-polymer conjugate can be in the form of a flat sheet, powder, large-sized particle, fiber, etc.

A most distinguishing feature of this method for immobilizing enzymes is that it is a *chemical* method. The immobilization is dependent on the formation of covalent bonds between the enzyme and the support, and a derivatized enzyme is produced. Consequently, because of a chemical alteration in its structure, the derivatized enzyme *may* exhibit superior chemical or physical properties relative to its soluble counterpart.

Another advantage of the method is its high degree of development. The preparation and characterization of enzyme-polymer conjugates have been described quite adequately and the method gives an individual a wide choice of supports for immobilizing enzymes.

Further, certain enzyme-polymer conjugates described in the literature and given in Table 4 are readily available. A list of some commercially available enzyme-polymer conjugates is given in Table 5. This list is by no means complete, and is intended to serve mainly as a guide to the suppliers of the derivatized enzymes and to show what supports have been used. Many of the supports mentioned are commercially available.

There are certainly some disadvantages to this method of enzyme immobilization. In preparing covalently linked water-insoluble enzyme polymer conjugates, it is highly desirable to know something about the chemical make-up or requirements of the enzyme. It is extremely helpful, if not essential, to know at least the types of amino acid residues that are essential for catalysis and binding. A "shot-in-the-dark" approach can be used, but this often leads to complete inactivation of the catalyst.

Another disadvantage of the method is that occasionally the successful attachment of an enzyme to a support requires elaborate preparations of both the enzyme and the support. Certain enzymes are extremely sensitive to changes in pH, ionic strength, etc., and the conditions necessary for their successful immobilization can completely abolish activity.

This chemical method can give derivatives of unsuitable physical character, can exhibit low efficiency in the coupling procedure, and can give derivatives of low catalytic efficiency. A serious disadvantage of water-insoluble, enzyme-polymer conjugates is their low catalytic efficiency on high molecular weight substrates, caused mainly by steric repulsion of the macromolecules. This aspect is discussed further in Chapter 3.

TABLE 4

**Covalently Bonded Water-Insoluble Enzyme Derivatives**

| Enzyme immobilized | Parent polymer | Activation procedure | Functional group of polymer involved in covalent bonding with enzyme | Ref. |
|---|---|---|---|---|
| Alcohol dehydrogenase (1.1.1.1) | Methacrylic acid and methacrylic acid-*m*-fluoroanilide copolymer (crosslinked with divinylbenzene) | Nitric acid | Aryl fluoride | 41, 110, 111 |
| | Polymethacrylic acid anhydride (crosslinked with 1,6-diaminohexane) | None | Anhydride | 92 |
| Lactate dehydrogenase (1.1.1.27) | Brick | Sulfuryl chloride | (Silyl chloride?) | 112 |
| | Brick | Thionyl chloride | (Silyl chloride?) | 112 |
| | Cellulose, diethylaminoethyl ether | Cyanuric chloride | Chloride | 40 |
| | Cellulose, diethylaminoethyl ether | Procion® brilliant orange MGS (dichloro-*s*-triazine) | Chloride | 62, 63 |
| | Polymethacrylic acid anhydride (crosslinked with 1,6-diaminohexane) | None | Anhydride | 92 |
| Glucose-6-phosphate dehydrogenase (1.1.1.49) | Acrylamide and acrylic acid copolymer (crosslinked with *N,N'*-methylenebisacrylamide); combination method–simultaneous covalent attachment and entrapment | CMC | Carboxyl | 107 |
| Glucose-6-phosphate dehydrogenase (1.1.1.49) – hexokinase (2.7.1.1) | Acrylamide and acrylic acid copolymer (crosslinked with *N,N'*-methylenebisacrylamide) | CMC | Carboxyl | 107 |
| | Sepharose® | Cyanogen bromide | Imidocarbonate | 107 |
| Glucose oxidase (1.1.3.4) | Polyacrylamide | Glutaraldehyde | Carbonyl | 78 |
| | Alumina (aminoalkyl derivative?) | Thiophosgene | Isothiocyanato | 113 |
| | Cellulose | Transition metal salts ($TiCl_4$) | Transition metal | 83 |
| | Cellulose, carboxymethyl ether | DCC | Carboxyl | 114 |
| | Cellulose, carboxymethyl ether, hydrazide | Nitrous acid | Acyl azide | 113 |
| | Glass, aminoaryl derivative | Nitrous acid | Diazonium | 115 |

TABLE 4 (Continued)

**Covalently Bonded Water-Insoluble Enzyme Derivatives**

| Enzyme immobilized | Parent polymer | Activation procedure | Functional group of polymer involved in covalent bonding with enzyme | Ref. |
|---|---|---|---|---|
| Glucose oxidase (1.1.3.4) | Glass, aminoaryl derivative | Thiophosgene | Isothiocyanato | 113 |
| | Hydroxyapatite, (aminoalkyl derivative?) | Thiophosgene | Isothiocyanato | 113 |
| | Nickel oxide, aminoalkyl derivative | Thiophosgene | Isothiocyanato | 104 |
| Glyceraldehydephosphate dehydrogenase (1.2.1.12) | Cellulose, aminoethyl ether | Glutaraldehyde | Carbonyl | 116, 117 |
| Luciferase (1.2.-.-) | Polyacrylic acid | Methanol, hydrazine, nitrous acid | Acyl azide | 118 |
| Glutamate dehydrogenase (1.4.1.3) | Cellulose, carboxymethyl ether | EDAPC | Carboxyl | 119 |
| | Collagen | Hydrazine, nitrous acid | Acyl azide | 105 |
| | Sephadex®, carboxymethyl ether (C-50) | EDAPC | Carboxyl | 119 |
| | Bio-Gel® P-150, carboxyl[a] | EDAPC | Carboxyl | 119 |
| | Amberlite® CG-50[a] | EDAPC | Carboxyl | 119 |
| L-Amino acid oxidase (1.4.3.2) | Glass, aminoaryl derivative | Nitrous acid | Diazonium | 120, 121 |
| Tyrosinase (1.10.3.1) | Cellulose, diethylaminoethyl ether | 2-amino-4, 6-dichloro-*s*-triazine | Chloride | 122 |
| Catalase (1.11.1.6) | Polyaminostyrene | Phosgene | Isocyanato | 37, 38 |
| | Cellulose | Cyanuric chloride | Chloride | 123, 124 |
| | Cellulose, *p*-aminobenzyl ether | Nitrous acid | Diazonium | 123 |
| | Cellulose, 3-chloro-2-hydroxypropyl ether | None (?) | Alkyl chloride (?) | 124 |
| | Cellulose, 4-$\beta$-hydroxyethylsulfanil-2-amino-anisole ether | Nitrous acid (?) | Diazonium (?) | 124 |
| | Cheesecloth (cotton) | Sodium periodate | Carbonyl | 100 |
| | Poliodal-4 | None | Alkyl iodide | 80 |

TABLE 4 (Continued)

**Covalently Bonded Water-Insoluble Enzyme Derivatives**

| Enzyme immobilized | Parent polymer | Activation procedure | Functional group of polymer involved in covalent bonding with enzyme | Ref. |
|---|---|---|---|---|
| Peroxidase (1.11.1.7) | Cellulose, aminoethyl ether | DCC | Amino | 125 |
| | Cellulose, carboxymethyl ether | DCC | Carboxyl | 114, 125, 126 |
| | Cellulose, carboxymethyl ether, benzidine derivative | Nitrous acid | Diazonium | 114 |
| Hexokinase (2.7.1.1) | Cellulose (cellophane) | Cyanuric chloride | Chloride | 99 |
| | Sephadex | Cyanogen bromide | Imidocarbonate | 127 |
| Hexokinase (adsorbed on colloidal silica, crosslinked with glutaraldehyde) | Cellulose, aminoethyl ether | 1-(*p*-nitrophenyl)-4-(*p*-methylthiophenyl) succinate, oxidation | Ester | 108 |
| Pyruvate kinase (2.7.1.40) | Cellulose | Cyanuric chloride | Chloride | 40, 62 |
| | Cellulose, diethylaminoethyl ether[a] | Cyanuric chloride | Chloride | 40 |
| Creatine kinase (2.7.3.2) | Cellulose, *p*-aminobenzyl ether | Nitrous acid | Diazonium | 128 |
| | Cellulose, carboxymethyl ether, hydrazide | Nitrous acid | Acyl azide | 128 |
| | Cellulose, diethylaminoethyl ether | 2,4-dichloro-6-carboxymethylamino-*s*-triazine | Chloride | 62 |
| Polynucleotide phosphorylase (2.7.7.8) | Cellulose | Cyanogen bromide | Imidocarbonate | 129 |
| Ribonuclease A (2.7.7.16) | Polyacrylamide | Glutaraldehyde | Carbonyl | 78 |
| | Polyaminostyrene | Nitrous acid | Diazonium | 35, 36 |
| | Bentonite | Cyanuric chloride | Chloride | 130 |
| | Cellulose, aminobenzoyl ester | Nitrous acid | Diazonium | 131 |

TABLE 4 (Continued)

**Covalently Bonded Water-Insoluble Enzyme Derivatives**

| Enzyme immobilized | Parent polymer | Activation procedure | Functional group of polymer involved in covalent bonding with enzyme | Ref. |
|---|---|---|---|---|
| Ribonuclease A (2.7.7.16) | Cellulose, *p*-aminobenzyl ether | Nitrous acid | Diazonium | 57, 123 |
| | Cellulose, carboxymethyl ether, hydrazide | Nitrous acid | Acyl azide | 131, 132 |
| | Cellulose, carboxymethyl ether | DCC | Carboxyl | 114 |
| | Sepharose | Cyanogen bromide | Imidocarbonate | 133 |
| | Sepharose (crosslinked with epichlorohydrin) | Cyanogen bromide | Imidocarbonate | 134 |
| Ribonuclease (polytyrosyl derivative) | *p*-amino-DL-Phe and L-Leu copolymer | Nitrous acid | Diazonium | 53 |
| tRNA Nucleotidyltransferase (2.7.7.20) | Sepharose | Cyanogen bromide | Imidocarbonate | 135 |
| Ribonuclease $T_1$ (2.7.7.26) | Cellulose, *p*-aminobenzyl ether | Nitrous acid | Diazonium | 136, 137 |
| | Cellulose, carboxymethyl ether, (hydrazide?) | Nitrous acid | Acyl azide | 136 |
| | Sephadex | Cyanogen bromide | Imidocarbonate | 137 |
| | Sepharose | Cyanogen bromide | Imidocarbonate | 137 |
| Lipase (3.1.1.3) | Polyaminostyrene | Phosgene | Isocyanato | 38 |
| Acetylcholinesterase (3.1.1.7) | Glass, aminoalkyl derivative | CMC | Amino | 138 |
| | Glass, aminoaryl derivative | Nitrous acid | Diazonium | 138 |
| | Sepharose | Cyanogen bromide | Imidocarbonate | 139 |
| Cholinesterase (3.1.1.8) | L-Ala and L-Glu copolymer | WRK | Carboxyl | 140 |
| | Poly-Asp | WRK | Carboxyl | 140 |
| | Cellulose, carboxymethyl ether | WRK | Carboxyl | 140 |
| | Cellulose, diethylaminoethyl ether | Procion brilliant orange | Chloride | 96 |
| | Polygalacturonic acid | WRK | Carboxyl | 140 |
| | Sepharose | Cyanogen bromide | Imidocarbonate | 139 |

TABLE 4 (Continued)

**Covalently Bonded Water-Insoluble Enzyme Derivatives**

| Enzyme immobilized | Parent polymer | Activation procedure | Functional group of polymer involved in covalent bonding with enzyme | Ref. |
|---|---|---|---|---|
| Steroid esterase (3.1.1.-) | Glass, aminoaryl derivative | Nitrous acid | Diazonium | 141 |
| Alkaline phosphatase (3.1.3.1) | Brick | Sulfuryl chloride | (Silyl chloride?) | 112 |
| | Brick | Thionyl chloride | (Silyl chloride?) | 112 |
| | Cellulose, carboxymethyl ether | DCC | Carboxyl | 114 |
| | Glass, aminoaryl derivative | Nitrous acid | Diazonium | 39, 60 |
| | Maleic anhydride and ethylene copolymer | None | Anhydride | 39 |
| Alkaline phosphatase (3.1.3.1) | Maleic anhydride and methyl vinyl ether copolymer (Gantrez® AN 169) | None | Anhydride | 39 |
| | Methacrylic acid and methacrylic acid-*m*-fluoroanilide copolymer (linear and crosslinked) | Nitric acid | Aryl fluoride | 39 |
| Acid phosphatase (3.1.3.2) | Polyacrylamide | Glutaraldehyde | Carbonyl | 78 |
| | Cellulose, carboxymethyl ether | DCC | Carboxyl | 114 |
| | Polymethacrylic acid anhydride (crosslinked with 1,6-diaminohexane) | None | Anhydride | 92 |
| Hexosediphosphatase (fructose-1,6-diphosphatase) (3.1.3.11) | Cellulose, aminoethyl ether | Glutaraldehyde | Carbonyl | 116, 117 |
| Deoxyribonuclease (3.1.4.5) | L-Ala and L-Glu copolymer | WRK | Carboxyl | 140 |
| | Cellulose, carboxymethyl ether | DCC | Carboxyl | 114 |
| | Cellulose, carboxymethyl ether | WRK | Carboxyl | 140 |
| | Cellulose, carboxymethyl ether, hydrazide | Nitrous acid | Acyl azide | 140 |
| | Polygalacturonic acid | WRK | Carboxyl | 140 |
| | Maleic anhydride and ethylene copolymer | None | Anhydride | 140 |
| Deoxyribonuclease I (3.1.4.5) | Glass, aminoaryl derivative | Nitrous acid | Diazonium | 142 |

TABLE 4 (Continued)

**Covalently Bonded Water-Insoluble Enzyme Derivatives**

| Enzyme immobilized | Parent polymer | Activation procedure | Functional group of polymer involved in covalent bonding with enzyme | Ref. |
|---|---|---|---|---|
| Staphylococcal nuclease (3.1.4.7) | Sepharose | Cyanogen bromide | Imidocarbonate | 133, 143 |
| Staphylococcal nuclease (Fragment 6-48) | Sepharose | Cyanogen bromide | Imidocarbonate | 144 |
| Staphylococcal nuclease (Fragment 49-149) | Sepharose | Cyanogen bromide | Imidocarbonate | 143 |
| Sterol sulfatase (3.1.6.2) | Glass, aminoaryl derivative | Nitrous acid | Diazonium | 145 |
| α-Amylase (diastase) (3.2.1.1) | Polyaminostyrene | Nitrous acid | Diazonium | 35, 36, 146 |
| | Cellulose | Transition metal salts ($TiCl_4$) | Transition metal | 83 |
| | Cellulose, *p*-aminobenzyl ether | Nitrous acid | Diazonium | 146 |
| | Cellulose, 3-(*p*-aminophenoxy)-2-hydroxypropyl ether | Nitrous acid | Diazonium | 74, 147 |
| | Cellulose, 3-(*p*-aminophenoxy)-2-hydroxypropyl ether | Thiophosgene | Isothiocyanato | 74, 147 |
| | Cellulose, carboxymethyl ether, hydrazide | Nitrous acid | Acyl azide | 146, 146a |
| | Enzacryl® AA | Nitrous acid | Diazonium | 56, 73-75 |
| | Enzacryl AA | Thiophosgene | Isothiocyanato | 56, 73-75 |
| | Enzacryl AH | Nitrous acid | Acyl azide | 56, 73-75 |
| | Enzacryl Polyacetal | Acid | Carbonyl | 56, 76, 83 |
| | Methacrylic acid and methacrylic acid-*m*-flouroanilide copolymer (crosslinked with divinylbenzene) | Nitric acid | Aryl fluoride | 41, 110, 111 |
| | Methacrylic acid and 2,4-dinitro-5-fluorostyrene copolymer (crosslinked with divinylbenzene) | None | Aryl fluoride | 148 |
| | Methacrylic acid and 3,5-dinitro-4-fluorostyrene copolymer (crosslinked with divinylbenzene) | None | Aryl fluoride | 148 |
| | Polystyrene, *p*-fluorophenylsulfone | Sulfonation, nitration | Aryl fluoride | 148 |
| β-Amylase (3.2.1.2) | Bentonite | Cyanuric chloride | Chloride | 130 |
| | Cellulose, 3-(*p*-aminophenoxy)-2-hydroxypropyl ether | Nitrous acid | Diazonium | 42, 74 |

TABLE 4 (Continued)

**Covalently Bonded Water-Insoluble Enzyme Derivatives**

| Enzyme immobilized | Parent polymer | Activation procedure | Functional group of polymer involved in covalent bonding with enzyme | Ref. |
|---|---|---|---|---|
| β-Amylase (3.2.1.2) | Cellulose, 3-(*p*-aminophenoxy)-2-hydroxypropyl ether | Thiophosgene | Isothiocyanato | 42, 74 |
| | Enzacryl AA | Nitrous acid | Diazonium | 73, 74 |
| | Enzacryl AA | Thiophosgene | Isothiocyanato | 73, 74 |
| | Enzacryl AH[a] | Nitrous acid | Acyl azide | 73 |
| | Sephadex, 3-(*p*-aminophenoxy)-2-hydroxypropyl ether | Thiophosgene | Isothiocyanato | 149 |
| Glucoamylase (amyloglucosidase or γ-amylase) (3.2.1.3) | Cellulose | Transition metal salts ($TiCl_4$, $TiCl_3$, $SnCl_4$, $SnCl_2$ $ZrCl_4$, $VCl_3$, $FeCl_2$) | Transition metal | 83 |
| | Cellulose, 3-(*p*-aminophenoxy)-2-hydroxypropyl ether | Nitrous acid | Diazonium | 42 |
| | Cellulose, 3-(*p*-aminophenoxy)-2-hydroxypropyl ether[a] | Thiophosgene | Isothiocyanato | 42 |
| | Cellulose, carboxymethyl ether, hydrazide | Nitrous acid | Acyl azide | 150 |
| | Cellulose, diethylaminoethyl ether | 2-amino-4,6-dichloro-*s*-triazine | Chloride | 65, 151 |
| | Glass | Transition metal salts ($TiCl_4$, $TiCl_3$, $SnCl_4$) | Transition metal | 83 |
| | Nylon | Transition metal salt ($TiCl_4$) | Transition metal | 83 |
| | Poliodal-4 | None | Alkyl iodide | 80 |
| Cellulase (3.2.1.4) | Bentonite | Cyanuric chloride | Chloride | 130 |
| | Methacrylic acid and methacrylic acid-*m*-fluoroanilide copolymer (crosslinked with divinylbenzene) | Nitric acid | Aryl fluoride | 41 |
| Dextranase (3.2.1.11) | Enzacryl Polyacetal | Acid | Carbonyl | 56, 76, 83 |
| | Methacrylic acid and methacrylic acid-*m*-fluoro-anilide copolymer (crosslinked with divinylbenzene)[a] | Nitric acid | Aryl fluoride | 41 |

TABLE 4 (Continued)

**Covalently Bonded Water-Insoluble Enzyme Derivatives**

| Enzyme immobilized | Parent polymer | Activation procedure | Functional group of polymer involved in covalent bonding with enzyme | Ref. |
|---|---|---|---|---|
| β-Glucosidase (3.2.1.21) | Cellulose | Ethyl chloroformate-triethylamine | Cyclic carbonate | 152 |
| β-Galactosidase (3.2.1.23) | Cellulose | Cyanuric chloride | Chloride | 62, 64 |
| | Cellulose, diethylaminoethyl ether | 2,4-dichloro-6-carboxymethylamino-*s*-triazine | Chloride | 62, 64, 97 |
| | Glass, aminoalkyl derivative | Glutaraldehyde | Carbonyl | 153 |
| | Glass, aminoaryl derivative | Nitrous acid | Diazonium | 154 |
| β-Galactosidase (3.2.1.23) – hexokinase (2.7.1.1) – glucose-6-phosphate dehydrogenase (1.1.1.49) | Sephadex G-50C | Cyanogen bromide | Imidocarbonate | 109 |
| β-Fructofuranosidase (invertase) (3.2.1.26) | Polyaminostyrene | Nitrous acid | Diazonium | 102 |
| | Bentonite | Cyanuric chloride | Chloride | 130 |
| | Bentonite | Sulfuryl chloride | (Silyl chloride?) | 112 |
| | Bentonite | Thionyl chloride | (Silyl chloride?) | 112 |
| | Brick | Cyanuric chloride | Chloride | 112 |
| | Brick | Sulfuryl chloride | (Silyl chloride?) | 112 |
| | Brick | Thionyl chloride | (Silyl chloride?) | 112 |
| | Cellulose | Transition metal salts ($TiCl_4$) | Transition metal | 83 |
| | Methacrylic acid and methacrylic acid-*m*-fluoroanilide copolymer (crosslinked with divinylbenzene) | Nitric acid | Aryl fluoride | 41, 67, 110 |
| | Glass | Cyanuric chloride | Chloride | 112 |
| | Glass | Sulfuryl chloride | (Silyl chloride?) | 112 |
| | Glass | Thionyl chloride | (Silyl chloride?) | 112 |
| β-Glucuronidase (3.2.1.31) | Cellulose, carboxymethyl ether | DCC | Carboxyl | 114 |
| Hyaluronidase (3.2.3.35) | Agarose | 2-amino-4,6-dichloro-*s*-triazine | Chloride | 155 |

TABLE 4 (Continued)

**Covalently Bonded Water-Insoluble Enzyme Derivatives**

| Enzyme immobilized | Parent polymer | Activation procedure | Functional group of polymer involved in covalent bonding with enzyme | Ref. |
|---|---|---|---|---|
| Naringinase (3.2.-.-) | Maleic anhydride and ethylene copolymer | None | Anhydride | 156 |
| | Maleic anhydride and isolbutylvinyl ether copolymer | None | Anhydride | 156 |
| | Maleic anhydride and methylvinyl ether copolymer | None | Anhydride | 156 |
| | Maleic anhydride and styrene copolymer | None | Anhydride | 156 |
| | Maleic anhydride and vinyl pyrollidone copolymer[b] | None | Anhydride | 156 |
| | Cellulose, *p*-aminobenzyl ether[a] | Nitrous acid | Diazonium | 156 |
| | S-MDA[a] | Nitrous acid | Diazonium | 156 |
| Leucine aminopeptidase (3.4.1.1) | Sephadex | Cyanogen bromide | Imidocarbonate | 157 |
| Aminopeptidase-M (3.4.1.-) | Sepharose | Cyanogen bromide | Imidocarbonate | 158 |
| Carboxypeptidase A (3.4.2.1) | Polyaminostyrene | Nitrous acid | Diazonium | 35, 36 |
| Carboxypeptidase B (3.4.2.2) | *p*-amino-DL-Phe and L-Leu copolymer | Nitrous acid | Diazonium | 159 |
| | Carbohydrate complex (unknown structure) | (Cyanogen bromide?) | (Imidocarbonate) | 159 |
| | Maleic anhydride and ethylene copolymer | None | Anhydride | 159, 160 |
| | Sepharose | Cyanogen bromide | Imidocarbonate | 159 |
| Carboxypeptidase N (3.4.2.-) | Maleic anhydride and ethylene copolymer | None | Anhydride | 159 |
| Prolidase (3.4.3.7) | Sepharose | Cyanogen bromide | Imidocarbonate | 158 |
| Pepsin (3.4.4.1) | Polyaminostyrene | Nitrous acid | Diazonium | 35, 36 |
| | Glass, aminoalkyl derivative | CMC | Amino | 161 |

TABLE 4 (Continued)

**Covalently Bonded Water-Insoluble Enzyme Derivatives**

| Enzyme immobilized | Parent polymer | Activation procedure | Functional group of polymer involved in covalent bonding with enzyme | Ref. |
|---|---|---|---|---|
| Pepsin (3.4.4.1) | Glass | ? | ? | 162 |
| | Methacrylic acid and methacrylic acid-*m*-fluororanilide copolymer (crosslinked with divinylbenzene) | Nitric acid | Aryl fluoride | 41, 110, 111 |
| | Sepharose | Cyanogen bromide *p*-phenylenediamine, acetaldehyde, 3-dimethylaminopropyl isocyanide (Ugi reaction) | Amino | 86 |
| | Sepharose | Cyanogen bromide, 4,4′-methylenedianiline, acetaldehyde, 3-dimethyl-propyl isocyanide (Ugi reaction) | Amino | 86 |
| | Sepharose (crosslinked) | Cyanogen bromide, *p*-phenylenediamine, acetaldehyde, cyclohexyl isocyanide (Ugi reaction) | Amino | 86 |
| Rennin (3.4.4.3) | Cellulose, aminoethyl | Glutaraldehyde | Carbonyl | 44 |
| | Sepharose | Cyanogen bromide | Imidocarbonate | 44 |
| Trypsin (3.4.4.4) | Polyacrylamide | Hydrazine, nitrous acid | Acyl azide | 77, 163 |
| | Polyacrylamide | Glutaraldehyde | Carbonyl | 78 |
| | Acrylamide and acrylic acid copolymer | CMC | Carboxyl | 106 |
| | Acrylamide and acrylic acid copolymer (crosslinked with *N,N′*-methylenebisacrylamide); combination of methods–simultaneous covalent bond formation and entrapment | CMC | Carboxyl | 106 |
| | Acrylamide and hydroxyethyl methacrylate co-polymer | Cyanogen bromide | Imidocarbonate | 106 |
| | Acrylamide and methyl acrylate copolymer | Hydrazine, nitrous acid | Acyl azide | 164 |

TABLE 4 (Continued)

**Covalently Bonded Water-Insoluble Enzyme Derivatives**

| Enzyme immobilized | Parent polymer | Activation procedure | Functional group of polymer involved in covalent bonding with enzyme | Ref. |
|---|---|---|---|---|
| Trypsin (3.4.4.4) | L-Ala and L-Glu copolymer | WRK | Carboxyl | 165 |
| | *p*-amino-DL-Phe and L-Leu copolymer | Nitrous acid | Diazonium | 55 |
| | Bentonite | Cyanuric chloride | Chloride | 130 |
| | Cellulose | Cyanogen bromide | Imidocarbonate | 48 |
| | Cellulose | Cyanuric chloride | Chloride | 123 |
| | Cellulose | Transition metal salts ($TiCl_4$) | Transition metal | 83 |
| | Cellulose, *p*-aminobenzoyl ester | Nitrous acid | Diazonium | 123 |
| | Cellulose, *p*-aminobenzyl ether | Nitrous acid | Diazonium | 47, 123 |
| | Cellulose, *p*-aminobenzyl ether | Glutaraldehyde | Carbonyl | 45 |
| | Cellulose, *m*-aminobenzyloxymethyl ether | Nitrous acid | Diazonium | 123 |
| | Cellulose, aminoethyl ether | Glutaraldehyde | Carbonyl | 45, 45a, 166 |
| | Cellulose, 3-amino-4-methoxyphenylsulfonylethyl ether | Nitrous acid | Diazonium | 123 |
| | Cellulose, carboxymethyl ether | DCC | Carboxyl | 114 |
| | Cellulose, carboxymethyl ether, hydrazide | Nitrous acid | Acyl azide | 57, 113, 132 167, 168 |
| | Cellulose, active golden yellow KX derivative | Nitrous acid | Diazonium | 123 |
| | Enzacryl Polyacetal | Acid | Carbonyl | 56, 76 |
| | Glass, aminoalkyl derivative | Thiophosgene | Isothiocyanato | 59, 113 |
| | Glass aminoaryl derivative | Nitrous acid | Diazonium | 59, 113 |
| | Maleic anhydride and ethylene copolymer | None | Anhydride | 54, 68-70, 113 159, 160, 169-181 |
| | Polymethacrylic acid anhydride (crosslinked with 1,6-diaminohexane) | None | Anhydride | 92 |
| | Methacrylic acid and methacrylic acid-*m*-fluoroanilide copolymer (crosslinked with divinylbenzene)[a] | Nitric acid | Aryl fluoride | 41 |
| | Nylon | Acid hydrolysis, nitrous acid, benzidine-DCC, nitrous acid | Diazonium | 101 |
| | Nylon | Acid hydrolysis, nitrous acid, hydrazine-DCC, nitrous acid | Acyl azide | 101 |

TABLE 4 (Continued)

**Covalently Bonded Water-Insoluble Enzyme Derivatives**

| Enzyme immobilized | Parent polymer | Activation procedure | Functional group of polymer involved in covalent bonding with enzyme | Ref. |
|---|---|---|---|---|
| Trypsin (3.4.4.4) | Poliodal-4 | None | Alkyl iodide | 82 |
| | Copoliodal-4 | None | Alkyl iodide | 82, 182 |
| | Sephadex | Cyanogen bromide | Imidocarbonate | 48, 183, 184 |
| | Sephadex, 3-*p*-aminophenoxy-2-hydroxypropyl ether | Thiophosgene | Isothiocyanato | 149 |
| | Sepharose | Cyanogen bromide | Imidocarbonate | 48, 158, 159, 183-185 |
| Trypsin (acetyl derivative) | *p*-amino-DL-Phe and L-Leu copolymer | Nitrous acid | Diazonium | 55 |
| Trypsin (*N*-acetylhomocysteine thiolactone derivative) | Enzacryl Polythiol | Potassium ferricyanide | Sulfhydryl | 56 |
| Trypsin (polytyrosyl derivative) | *p*-amino-DL-Phe and L-Leu copolymer | Nitrous acid | Diazonium | 52-55, 68, 169, 170, 172 |
| | S-MDA | Nitrous acid | Diazonium | 47 |
| Trypsinogen | Maleic anhydride and ethylene copolymer | None | Anhydride | 186 |
| | Sepharose | Cyanogen bromide | Imidocarbonate | 187 |
| α-Chymotrypsin (3.4.4.5) | Polyacrylamide | Glutaraldehyde | Carbonyl | 78 |
| | Polyacrylamide, aminoethyl derivative | Ugi reaction | Amino | 85 |
| | Polyacrylic acid | WRK | Carboxyl | 94 |
| | Agarose, carboxymethyl ether | Ugi reaction | Carboxyl | 85 |
| | L-Ala and L-Glu copolymer | WRK | Carboxyl | 165 |
| | Amberlite CG-50[a] | EDAPC | Carboxyl | 119 |
| | Bio-Gel P-150-carboxyl | EDAPC | Carboxyl | 119 |
| | Cellulose | Cyanogen bromide | Imidocarbonate | 48 |
| | Cellulose | Cyanuric chloride | Chloride | 61, 62, 123 |
| | Cellulose | 2-amino-4,6-dichloro-*s*-triazine | Chloride | 188 |
| | Cellulose | 2,4-dichloro-6-carboxymethylamino-*s*-triazine | Chloride | 61 |

TABLE 4 (Continued)

**Covalently Bonded Water-Insoluble Enzyme Derivatives**

| Enzyme immobilized | Parent polymer | Activation procedure | Functional group of polymer involved in covalent bonding with enzyme | Ref. |
|---|---|---|---|---|
| α-Chymotrypsin (3.4.4.5) | Cellulose | 2,4-dichloro-6-carboxy-methyloxy ether | Chloride | 61 |
| | Cellulose, *p*-aminobenzoyl ester | Nitrous acid | Diazonium | 123, 131 |
| | Cellulose, *p*-aminobenzyl ether | Nitrous acid | Diazonium | 57, 123 |
| | Cellulose, 3-aminobenzyloxymethyl ether | Nitrous acid | Diazonium | 123 |
| | Cellulose, 3-amino-4-methoxyphenylsulfonylethyl ether | Nitrous acid | Diazonium | 123 |
| | Cellulose, carboxymethyl ether | DCC | Carboxyl | 114 |
| | Cellulose, carboxymethyl ether | EDAPC | Carboxyl | 119 |
| | Cellulose, carboxymethyl ether | WRK | Carboxyl | 94 |
| | Cellulose, carboxymethyl ether | 2-amino-4,6-dichloro-*s*-triazine | Chloride | 188 |
| | Cellulose, carboxymethyl ether, hydrazide | Nitrous acid | Acyl azide | 57, 128, 131, 189-191 |
| | Cellulose, diethylaminoethyl ether | 2-amino-4,6-dichloro-*s*-triazine | Chloride | 91, 188 |
| | Enzacryl AA | Ugi reaction | Amino | 85 |
| | Glass, aminoalkyl derivative | Glutaraldehyde | Carbonyl | 153 |
| | Poly-L-Glu | WRK | Carboxyl | 94 |
| | Keratin | Ugi reaction | Amino, carboxyl | 85 |
| | Maleic anhydride and ethylene copolymer | None | Anhydride | 70, 94, 174, 177, 180 |
| | Polymethacrylic acid anhydride (crosslinked with 1,6-diaminohexane) | None | Anhydride | 92 |
| | Poliodal-2 | None | Alkyl iodide | 43, 80 |
| | Poliodal-4 | None | Alkyl iodide | 43, 80 |
| | Copoliodal-2 | None | Alkyl iodide | 43 |
| | Copoliodal-4 | None | Alkyl iodide | 43 |
| | Sephadex | 2-amino-4,6-dichloro-*s*-triazine | Chloride | 188 |
| | Sephadex | Cyanogen bromide | Imidocarbonate | 48, 71, 95, 184 |

TABLE 4 (Continued)

**Covalently Bonded Water-Insoluble Enzyme Derivatives**

| Enzyme immobilized | Parent polymer | Activation procedure | Functional group of polymer involved in covalent bonding with enzyme | Ref. |
|---|---|---|---|---|
| α-Chymotrypsin (3.4.4.5) | Sephadex, 3-*p*-aminophenoxy-2-hydroxypropyl ether | Thiophosgene | Isothiocyanato | 149 |
| | Sephadex, carboxymethyl ether (C-50) | EDAPC | Carboxyl | 119 |
| | Sephadex, carboxymethyl ether (C-50) | Ugi reaction | Carboxyl | 85 |
| | Sepharose | 2-amino-4,6-dichloro-*s*-triazine | Chloride | 188 |
| | Sepharose | Cyanogen bromide | Imidocarbonate | 44, 48, 72, 158, 184, 192-194 |
| | Sepharose | Cyanogen iodide | Imidocarbonate | 48 |
| | Sepharose | Sodium periodate, Ugi reaction | Carbonyl | 85 |
| | Sepharose, lysine derivative | Ugi reaction | Amino | 85 |
| | Sepharose-papain conjugate | Ugi reaction | Amino, carboxyl | 85 |
| | Wool | Ugi reaction | Amino, carboxyl | 85 |
| α-Chymotrypsin (polytyrosyl derivative) | *p*-amino-DL-Phe and L-Leu copolymer | Nitrous acid | Diazonium | 53 |
| | polycationic polymer (?) | Nitrous acid | Diazonium | 53a |
| δ-Chymotrypsin | Poliodal-4 | None | Alkyl iodide | 43, 80 |
| Chymotrypsinogen | Poliodal-4 | None | Alkyl iodide | 43, 80 |
| Papain (3.4.4.10) | *p*-amino-DL-Phe and L-Leu copolymer | Nitrous acid | Diazonium | 113, 195-200 |
| | Cellulose, *p*-aminobenzyl | Nitrous acid | Diazonium | 47, 201 |
| | Cellulose, carboxymethyl ether, hydrazide | Nitrous acid | Acyl azide | 113, 202 |
| | Collagen | bisdiazobenzidine-3,3′-dicarboxylic acid | Diazonium | 200 |
| | Collagen | bisdiazobenzidine-2,2′-disulfonic acid | Diazonium | 200 |
| | Enzacryl Polyacetal | Acid | Carbonyl | 56, 76 |
| | Glass, aminoalkyl derivative (?) | Thiophosgene | Isothiocyanato | 113 |
| | Glass, aminoaryl derivative | Nitrous acid | Diazonium | 59, 113 |
| | Lewatit CA 9119 (poly-*p*-aminostyrene) | Thiophosgene | Isothiocyanato | 203, 204 |

TABLE 4 (Continued)

**Covalently Bonded Water-Insoluble Enzyme Derivatives**

| Enzyme immobilized | Parent polymer | Activation procedure | Functional group of polymer involved in covalent bonding with enzyme | Ref. |
|---|---|---|---|---|
| Papain (3.4.4.10) | Polymethacrylic acid anhydride (crosslinked with 1,6-diaminohexane) | None | Anhydride | 92 |
| | Methacrylic acid and methacrylic acid-*m*-fluoroanilide copolymer (crosslinked with divinylbenzene) | Nitric acid | Aryl fluoride | 204 |
| | Methacrylic acid and 2-fluorostyrene copolymer | Nitric acid | Aryl fluoride | 204 |
| | Methacrylic acid and 3-fluorostyrene copolymer | Nitric acid | Aryl fluoride | 148, 204 |
| | Methacrylic acid and 4-fluorostyrene copolymer | Nitric acid | Aryl fluoride | 148, 204 |
| | Sepharose | Cyanogen bromide | Imidocarbonate | 48, 85 |
| | S-MDA | Nitrous acid | Diazonium | 47 |
| | Polystyrene | 4-fluorophenyl-sulfonyl chloride, sulfonation, nitration | Aryl fluoride | 148 |
| | Polystyrene | Sulfonation, nitration, reduction, thiophosgene | Isothiocyanato | 203 |
| | Polyaminostyrene | Sulfonation, thiophosgene | Isothiocyanato | 204 |
| | *m*-Aminostyrene and methacrylic acid copolymer | Thiophosgene | Isothiocyanato | 203, 204 |
| | *m*-Isothiocyanatostyrene and methacrylic acid copolymer | None | Isothiocyanato | 203, 204 |
| | *m*-Isothiocyanatostyrene and acrylic acid copolymer | None | Isothiocyanato | 203, 204 |
| | Polyvinylamine | Thiophosgene | Isothiocyanato | 203, 204 |
| Papain (mercuribenzoate derivative) | Cellulose, *p*-aminobenzyl ether | Nitrous acid | Diazonium | 47 |
| | Sepharose | Cyanogen bromide | Imidocarbonate | 48 |
| | S-MDA | Nitrous acid | Diazonium | 47 |
| Ficin (3.4.4.12) | Cellulose, carboxymethyl ether | DCC | Carboxyl | 114 |
| | Cellulose, carboxymethyl ether, hydrazide | Nitrous acid | Acyl azide | 49, 113, 128, 131, 205 |
| | Glass, aminoalkyl derivative (?) | Thiophosgene | Isothiocyanato | 113 |
| | Glass, aminoaryl derivative | Nitrous acid | Diazonium | 113 |

TABLE 4 (Continued)

**Covalently Bonded Water-Insoluble Enzyme Derivatives**

| Enzyme immobilized | Parent polymer | Activation procedure | Functional group of polymer involved in covalent bonding with enzyme | Ref. |
|---|---|---|---|---|
| Ficin (mercuribenzoate derivative) (3.4.4.12) | Cellulose, carboxymethyl ether, hydrazide | Nitrous acid | Acyl azide | 49 |
| Thrombin (3.4.4.13) | *p*-amino-DL-Phe and L-Leu copolymer | Nitrous acid | Diazonium | 170, 206 |
| | Cellulose, *m*-aminobenzyloxymethyl ether | Nitrous acid | Diazonium | 207 |
| | Cellulose, bromoacetyl ester | None | Alkyl bromide | 208 |
| | Maleic anhydride and ethylene copolymer | None | Anhydride | 209 |
| Prothrombin | *p*-amino-DL-Phe and L-Leu copolymer | Nitrous acid | Diazonium | 54, 186, 206 |
| Prothrombin (polytyrosyl derivative) | *p*-amino-DL-Phe and L-Leu copolymer | Nitrous acid | Diazonium | 54 |
| Plasminogen (3.4.4.14; plasmin) | *p*-amino-DL-Phe and L-Leu copolymer | Nitrous acid | Diazonium | 210 |
| Renin (3.4.4.15) | Sepharose | Cyanogen bromide | Imidocarbonate | 159, 185 |
| Subtilopeptidase A (subtilisin Carlsberg) (3.4.4.16) | Cellulose, *p*-aminobenzyl ether | Nitrous acid | Diazonium | 47 |
| | Cellulose, carboxymethyl ether | DCC | Carboxyl | 114 |
| | S-MDA | Nitrous acid | Diazonium | 47 |
| Subtilopeptidase C (subtilisin BPN′ or Nagarse) | Cellulose, carboxymethyl ether | DCC | Carboxyl | 114 |
| Kallikrein (3.4.4.21) | Maleic anhydride and ethylene copolymer | None | Anhydride | 70, 159, 174, 177, 180 |
| | Sepharose | Cyanogen bromide | Imidocarbonate | 159 |
| Bromelain (3.4.4.24) | Cellulose, carboxymethyl ether, hydrazide | Nitrous acid | Acyl azide | 211, 212 |
| | Polymethacrylic acid anhydride (crosslinked with 1,6-diaminohexane) | None | Anhydride | 92 |

TABLE 4 (Continued)

**Covalently Bonded Water-Insoluble Enzyme Derivatives**

| Enzyme immobilized | Parent polymer | Activation procedure | Functional group of polymer involved in covalent bonding with enzyme | Ref. |
|---|---|---|---|---|
| Pronase (3.4.4.- and 3.4.1.-) | *p*-amino-DL-Phe and L-Leu copolymer | Nitrous acid | Diazonium | 213, 214 |
| | Cellulose, bromoacetyl ester | None | Alkyl bromide | 215 |
| | Glass, aminoaryl derivative | Nitrous acid | Diazonium | 216, 217 |
| | Glass, aminoalkyl derivative | Glutaraldehyde, polyethylenimine, glutaraldehyde | Carbonyl | 217 |
| | Polymethacrylic acid anhydride (crosslinked with 1,6-diaminohexane) | None | Anhydride | 92 |
| | Sephadex, carboxymethyl ether | EDAPC | Carboxyl | 215 |
| Streptokinase (3.4-.-?) | *p*-amino-DL-Phe and L-Leu copolymer | Nitrous acid | Diazonium | 218 |
| | Polyaminostyrene (insoluble?)[a] | Nitrous acid | Diazonium | 219 |
| | Cellulose, *p*-aminobenzyl ether | Nitrous acid | Diazonium | 220 |
| Pan protease (3.4.-.-) | Cellulose, carboxymethyl ether | DCC | Carboxyl | 114 |
| Proteinase (from *Arthrobacter*) (3.4.-.-) | Sepharose | Cyanogen bromide | Imidocarbonate | 221 |
| L-Asparaginase (3.5.1.1) | Cellulose | Cyanogen bromide | Imidocarbonate | 222 |
| | Cellulose, carboxymethyl ether | CMC | Carboxyl | 222 |
| | Cellulose, carboxymethyl ether | WRK | Carboxyl | 222 |
| | Dacron®, aminoaryl derivative | Nitrous acid | Diazonium | 103 |
| | Dextran, carboxymethyl ether, hydrazide | Nitrous acid | Acyl azide | 222 |
| Urease (3.5.1.5) | *p*-amino-DL-Phe andd L-Ala copolymer | Nitrous acid | Diazonium | 46 |
| | *p*-amino-DL-Phe and Gly copolymer | Nitrous acid | Diazonium | 46 |
| | *p*-amino-DL-Phe and L-Leu copolymer | Nitrous acid | Diazonium | 46 |
| | Bentonite | Cyanuric chloride | Chloride | 130 |

TABLE 4 (Continued)

**Covalently Bonded Water-Insoluble Enzyme Derivatives**

| Enzyme immobilized | Parent polymer | Activation procedure | Functional group of polymer involved in covalent bonding with enzyme | Ref. |
|---|---|---|---|---|
| Urease (3.5.1.5) | Enzacryl Polyacetal | Acid | Carbonyl | 56 |
| | Glass, aminoaryl derivative | Nitrous acid | Diazonium | 223 |
| | Nylon | Acid hydrolysis glutaraldehyde | Carbonyl | 224 |
| | Poliodal-4 | None | Alkyl iodide | 79, 82 |
| Urease (mercuribenzoate derivative) (3.5.1.5) | *p*-amino-DL-Phe and L-Leu copolymer | Nitrous acid | Diazonium | 46 |
| Penicillin amidase (3.5.1.11) | Cellulose | Cyanuric chloride | Chloride | 66 |
| | Cellulose, carboxymethyl ether, hydrazide | Nitrous acid | Acyl azide | 66 |
| | Cellulose, diethylaminoethyl ether | 2,4-dichloro-6-carboxymethylamino-*s*-triazine | Chloride | 66 |
| Aminoacylase (3.5.1.14) | Acrylamide and methyl acrylate copolymer (cross-linked with *N,N'*-methylenebisacrylamide) | Hydrazine, nitrous acid | Acyl azide | 164 |
| | Poly(methyl acrylate) (crosslinked with *N,N'*-methylenebisacrylamide) | Hydrazine, nitrous acid | Acyl azide | 164 |
| | Cellulose, *p*-aminobenzyl ether | Nitrous acid | Diazonium | 225 |
| | Cellulose, aminoethyl ether | DCC | Amino | 225 |
| | Cellulose, bromoacetyl ester | None | Alkyl bromide | 225 |
| | Cellulose, chloroacetyl ester | None | Alkyl chloride | 225 |
| | Cellulose, iodoacetyl ester | None | Alkyl iodide | 225 |
| | Cellulose, carboxymethyl ether | DCC | Carboxyl | 225 |
| | Cellulose, carboxymethyl ether, hydrazide | Nitrous acid | Acyl azide | 225 |
| | Sephadex | Cyanogen bromide | Imidocarbonate | 225 |
| ATPase (3.6.1.3) | Cellulose | Cyanuric chloride | Chloride | 98 |
| | Cellulose, carboxymethyl ether, hydrazide | None | Hydrazide (?) | 51, 226 |
| | Cellulose, carboxymethyl ether | None | Carboxyl (?) | 51 |
| | Cellulose, carboxymethyl ether, hydrazide | Nitrous acid | Acyl azide | 50, 51, 226, 227 |
| | Cellulose, carboxymethyl ether, hydrazide | Nitrous acid, re-arrangement | Isocyanato | 51 |

TABLE 4 (Continued)

**Covalently Bonded Water-Insoluble Enzyme Derivatives**

| Enzyme immobilized | Parent polymer | Activation procedure | Functional group of polymer involved in covalent bonding with enzyme | Ref. |
|---|---|---|---|---|
| Apyrase (3.6.1.5) | L-Ala and L-Glu copolymer | WRK | Carboxyl | 140 |
| | Polyaminostyrene | Nitrous acid | Diazonium | 93, 228 |
| | Poly-Asp | WRK | Carboxyl | 140 |
| | Cellulose | Cyanuric chloride | Chloride | 98 |
| | Cellulose, carboxymethyl ether | WRK | Carboxyl | 140 |
| | Cellulose, carboxymethyl ether, hydrazide | None | Hydrazide (?) | 51, 229 |
| | Cellulose, carboxymethyl ether, hydrazide | Nitrous acid | Acyl azide | 51, 93, 98, 228-230 |
| | Cellulose, carboxymethyl ether, hydrazide | Nitrous acid, rearrangement | Isocyanato | 51 |
| | Polygalacturonic acid | WRK | Carboxyl | 140 |
| | Maleic anhydride and ethylene copolymer | None | Anhydride | 93, 228 |
| | Poly(methyl methacrylate) | WRK | Carboxyl | 140 |
| Pyruvate decarboxylase (4.1.1.1) | Polyaminomethylstyrene (crosslinked with divinylbenzene) | CDAPC | Amino | 231 |
| | Polystyrene[c] | Cyanogen bromide | (?) | 231 |
| Fructosediphosphate aldolase (4.1.2.13) | Cellulose, *p*-aminobenzyl ether | (?) | (?) | 116 |
| | Cellulose, aminoethyl ether | Glutaraldehyde | Carbonyl | 116, 117 |
| | Maleic anhydride and ethylene copolymer | None | Anhydride | 116 |
| | Sepharose | Cyanogen bromide | Imidocarbonate | 232 |
| Threonine deaminase (4.2.1.16) | Brick | Sulfuryl chloride | (Silyl chloride?) | 112 |
| | Brick | Thionyl chloride | (Silyl chloride?) | 112 |
| Glucose isomerase (5.3.1.-) | Cellulose | Transition metal salt ($TiCl_4$) | Transition metal | 83 |
| Isoleucyl-tRNA synthetase (6.1.1.5) | Sepharose | Cyanogen bromide | Imidocarbonate | 233 |

[a]No activity observed
[b]Water-soluble derivative
[c]Unsuccessful attempt; not stated whether due to loss of activity or lack of coupling

TABLE 5

**Commercially Available Water-Insoluble, Polymer-Bound Enzymes**

| Immobilized enzyme | Matrix | Functional group of polymer used for immobilization | Supplier[a,b] |
|---|---|---|---|
| Alcohol dehydrogenase (1.1.1.1) | Cellulose, diethylaminoethyl ether | Triazinyl chloride | M |
| Glucose oxidase (1.1.3.4) | Cellulose, carboxymethyl ether | Acyl azide | GS, M |
| | Cellulose, diethylaminoethyl ether | Triazinyl chloride | M |
| | Sepharose | Imidocarbonate | W |
| Peroxidase (1.11.1.7) | Cellulose, carboxymethyl ether | Acyl azide | M |
| Ribonuclease A (2.7.7.16) | Cellulose, carboxymethyl ether | Acyl azide | EM, GS, M |
| | Maleic anhydride and divinyl ether copolymer | Anhydride | EM |
| | Sepharose | Imidocarbonate | W |
| Alkaline phosphatase (3.1.3.1) | Sepharose | Imidocarbonate | W |
| Acid phosphatase (3.1.3.2) | Polyacrylamide | Carbonyl | W |
| *α*-Amylase (3.2.1.1) | Cellulose, 3-(*p*-aminophenoxy)-2-hydroxypropyl ether | Diazonium | M |
| Leucine aminopeptidase (3.4.1.1) | Cellulose, diethylaminoethyl ether | Triazinyl chloride | M |
| Carboxypeptidase A (3.4.2.1) | Sepharose | Imidocarbonate | W |
| Trypsin (3.4.4.4) | Polyacrylamide | (?) | BMC |
| | Agarose | Triazinyl chloride | M |
| | Cellulose, carboxymethyl ether | Acyl azide | EM, GS, M |
| | Maleic anhydride and divinyl ether copolymer | Anhydride | EM |
| | Maleic anhydride and ethylene copolymer | Anhydride | M |
| | Sepharose | Imidocarbonate | W |
| Trypsin (polytyrosyl derivative) | S-MDA | Diazonium | M |
| *α*-Chymotrypsin (3.4.4.5) | Polyacrylamide | ? | BMC |
| | Agarose | Triazinyl chloride | M |
| | Cellulose, carboxymethyl ether | Acyl azide | EM, GS, M |
| | Maleic anhydride and ethylene copolymer | Anhydride | M |
| | Sepharose | Imidocarbonate | W |
| Papain (3.4.4.10) | Polyacrylamide | ? | BMC |

TABLE 5 (Continued)

**Commercially Available Water-Insoluble, Polymer-Bound Enzymes**

| Immobilized enzyme | Matrix | Functional group of polymer used for immobilization | Supplier[a,b] |
|---|---|---|---|
| Papain (3.4.4.10) | Agarose | Triazinyl chloride | M |
| | Cellulose, carboxymethyl ether | Acyl azide | EM, GS, M |
| | Maleic anhydride and ethylene copolymer | Anhydride | M |
| | Maleic anhydride and divinyl ether copolymer | Anhydride | EM |
| | S-MDA | Diazonium | M |
| Ficin (3.4.4.12) | Cellulose, carboxymethyl ether | Acyl azide | EM, GS, M |
| Subtilopeptidase A (3.4.4.16) | Maleic anhydride and ethylene copolymer | Anhydride | M |
| | S-MDA | Diazonium | M |
| Subtilopeptidase B (3.4.4.16) | Maleic anhydride and ethylene copolymer | Anhydride | M |
| | S-MDA | Diazonium | M |
| Subtilisin (not identified from *B. subtilis*) | Cellulose, carboxymethyl ether | Acyl azide | EM |
| Bromelain (3.4.4.24) | Cellulose, carboxymethyl ether | Acyl azide | EM, GS, M |
| Pronase (3.4.4.- and 3.4.1.-) | Cellulose, carboxymethyl ether | Acyl azide | EM |
| Protease (from *Streptomyces griseus*) (3.4.-.-) | Agarose | Triazinyl chloride | M |
| | Cellulose, carboxymethyl ether | Acyl azide | M |
| Proteinase K (3.4.-.-) | Cellulose, carboxymethyl ether | Acyl azide | EM |
| Urease (3.5.1.5) | Cellulose, diethylaminoethyl ether | Triazinyl chloride | M |

[a]Abbreviations used: BMC, Boehringer Mannheim Corp.; EM, EM Laboratories, Inc.; GS, Gallard-Schlesinger Chemical Mfg. Corp.; M, Miles Laboratories, Inc. (Research Division); W, Worthington Biochemical Corp.
[b]Trademarks associated with immobilized enzymes: Enzgel®, Boehringer Mannheim Corp.; Enzite®, Miles-Seravac (Pty) Ltd.

Chapter 3

# PROPERTIES OF COVALENTLY BONDED WATER-INSOLUBLE ENZYME-POLYMER CONJUGATES

A change in a chemical or physical property of an enzyme upon immobilization can be conveniently viewed as being due to either the nature of the water-insoluble support or to some alteration of the enzyme itself. Major factors of the water-insoluble support that can affect the properties of a covalently bonded enzyme are the diffusion layer surrounding the water-insoluble particle, steric repulsion of substrates, molecular size of the polymer, flexibility of the polymer backbone, degree of hydrophilicity, and electrostatic interactions. Factors pertaining to the chemically modified enzyme that can have a profound influence on particular properties are local and net charges, conformational changes, and transformations of catalytically essential amino acid residues. However, a resultant change in a specific property, be it favorable or unfavorable, may not be due to a single change in the enzyme. Any modification could cause a variety of distinct and mutually independent changes that finally manifest themselves as the "observed" or "resultant" effect. It is extremely difficult, if not impossible, to ascribe precisely the cause and magnitude of an alteration in a property of an enzyme upon its immobilization. Chemical and physical properties of covalently bonded, water-insoluble enzymes that have been extensively investigated are catalytic activity, pH-activity behavior, the Michaelis constant, substrate and inhibitor specificity, and various types of stabilities. A discussion of these properties follows.

**Activity**

The activity of a covalently bonded, water-insoluble, enzyme-polymer derivative can vary from nil[41,119,156] to apparently even higher[204] values than the native, water-soluble enzyme. In general, there seems to be no normal or expected range of activities for an immobilized enzyme, and the activity obtained is dependent on the particular enzyme, the specific support, and often upon the individual experimenter.

The usual way of expressing the activity of an enzyme is by specific activity, i.e., the number of μmoles of substrate transformed to product/min/mg of protein under certain specified conditions of temperature, pH, etc. Normally, the method is used also for expressing the activity of an immobilized enzyme. This means determining the number of μmoles of substrate transformed to product/min and the protein content of the enzyme-polymer conjugate. Quite often, only the number of units (μmoles of substrate transformed to product/min) per mg of *total* enzyme-polymer conjugate is given. Unfortunately, this manner of reporting activity does not allow a direct comparison of the efficiency of the immobilized enzyme with its freely soluble counterpart.

The extent of protein binding to a water-insoluble support varies according to the enzyme, the support, and the exact conditions employed during coupling. The amount of protein covalently bonded can be as low as several μg or as high as hundreds of mg/g of enzyme-polymer conjugate. Low levels of protein coupling (less than 10 mg/g of enzyme-polymer conjugate) were reported for β-glucosidase on cellulose carbonate,[152] trypsin on the crosslinked copolymer of acrylamide and methyl acrylate,[164] α-amylase on the isothiocyanato derivative of Enzacryl AA,[73] glucose oxidase on nickel oxide,[104] L-amino acid oxidase on the aminoaryl derivative of glass,[121] and α-chymotrypsin on 2-amino-4,6-dichloro-*s*-triazinyl derivatized cellulose.[188] High levels of protein binding (greater than 200 mg protein/g of enzyme-polymer conjugate) were reported for α-chymotrypsin on cyanogen bromide-activated Sepharose,[48] thrombin on *m*-aminobenzyloxymethylcellulose,[207] α-chymotrypsin on CM-Sephadex C-50 via the Ugi reaction,[85] polytyrosyl trypsin on S-MDA,[47] alkaline phosphatase on the copolymer of maleic anhydride and ethylene,[39] and α-chymotrypsin on 2-amino-4,6-dichloro-*s*-triazinyl derivatized Sepharose.[188] Exceptionally high protein bindings were reported for papain on

Goldstein et al.[69] expressed the qualitative considerations on the displacement of the pH-activity curves of immobilized enzymes mathematically. Assuming a Maxwell-Boltzmann distribution, the local hydrogen ion activity in the domain of the polyelectrolyte enzyme-polymer conjugate is given by Equation 41.

$$a^{I}_{H^+} = a^{II}_{H^+} \exp\left(\frac{z\epsilon\Psi}{kT}\right) \qquad (41)$$

Using the notation of Goldstein et al.,[69] $a^{I}_{H^+}$ and $a^{II}_{H^+}$ are the hydrogen ion activities in the polyelectrolyte-enzyme derivative phase (phase I) and the external solution (phase II), respectively, $\Psi$ is the electrostatic potential in the domain of the charged immobilized enzyme particle, $\epsilon$ is the positive electron charge, $z$ is a positive or negative integer (positive and one in the case of the hydrogen ion), k is the Boltzmann constant, and T is the absolute temperature. Equation 41 shows that the local hydrogen ion activity in the domain of a water-insoluble polyelectrolyte carrier is higher if the support is negatively charged and lower if it is positively charged. Since by definition $-\log a_{H^+} = \text{pH}$, Equation 41 can be rewritten as Equation 42,

$$\Delta\text{pH} = \text{pH}^{I} - \text{pH}^{II} = 0.43\,\frac{z\epsilon\Psi}{kT} \qquad (42)$$

which gives the pH difference between the local pH of the polyelectrolyte particle and the external solution.

Similar observations and conclusions were made by others using different enzymes and different negatively or positively charged carriers. For example, the following enzymes covalently bonded to negatively charged carriers exhibited a displacement of their pH-activity profile to more alkaline pH values: naringinase[156] on copolymers of maleic anhydride and ethylene or methylvinyl ether, trypsin[92] and chymotrypsin[92] on polymethacrylic acid anhydride, carboxypeptidase B[159] on the copolymer of maleic anhydride and ethylene, apyrase[140] on a copolymer of alanine and glutamic acid, CM-cellulose, methyl methacrylate, polygalacturonic acid, and polyaspartic acid, α-chymotrypsin[94] on polyacrylic acid, and ficin[49] on CM-cellulose. Conversely, amyloglucosidase,[151] penicillin amidase,[66] and α-chymotrypsin[188] bound to positively charged DEAE-cellulose exhibited a displacement of the pH-activity curves toward more acidic pH values. Partial or complete elimination of the displacement of the pH-activity curves with increasing ionic strength was reported for some but not all of these systems.[49,188]

The strongest evidence to date that the pH-activity curve of a covalently bonded enzyme can be shifted by the chemical modification was given by Axen et al.[95] and by Goldstein et al.[47] in studies with electrically neutral supports. Naturally, chemically induced effects can and do operate in highly charged carriers, but in these supports they are often masked by the considerably larger and stronger electrostatic effects induced by the polyelectrolytes. Axen et al.[95] observed that α-chymotrypsin bound to cyanogen bromide-activated Sephadex (initially a neutral carrier but made slightly electronegative by CNBr treatment) exhibited a considerably displaced pH–activity curve to more alkaline pH values. Increasing the ionic strength of the medium did not eliminate the displacement. Polytyrosyl trypsin, papain, and subtilopeptidase A covalently bound to diazotized S-MDA (a neutral carrier) similarly exhibited pH-activity profiles that were displaced toward alkaline pH's relative to the unmodified enzymes.[47] Again, the pH-activity curves of these immobilized enzymes were not affected by a change in the ionic strength of the medium. Figures 6 and 7 illustrate the pH-activity curves for

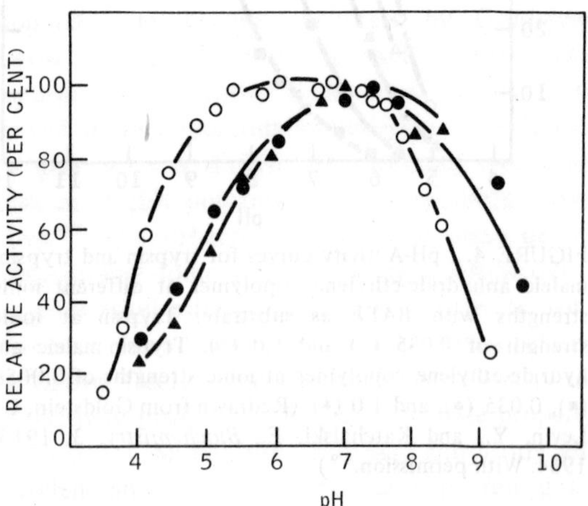


FIGURE 6. pH-Activity curves for papain and papain-S-MDA at different ionic strengths with BGEE as substrate. Papain at an ionic strength of 0.01 (○). Papain-S-MDA at ionic strengths of 0.01 (●) and 1.0 (▲). (Redrawn from Goldstein, L., Pecht, M., Blumberg, S., Atlas, D., and Levin, Y., *Biochemistry*, 9, 2322, 1970. With permission.[47])

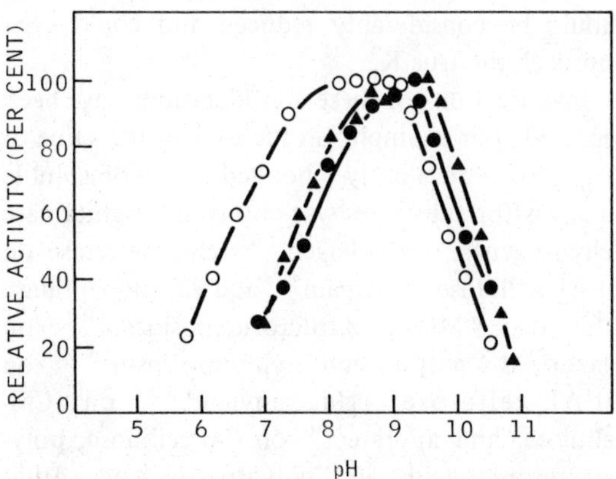


FIGURE 7. pH-Activity curves for subtilopeptidase A and subtilopeptidase A-S-MDA at different ionic strengths with ATEE as substrate. Subtilopeptidase A at an ionic strength of 0.01 (○); subtilopeptidase A-S-MDA at ionic strengths of 0.01 (●) and 0.55 (▲). (Redrawn from Goldstein, L., Pecht, M., Blumberg, S., Atlas, D., and Levin, Y., *Biochemistry,* 9, 2322, 1970. With permission.[47])

S-MDA immobilized papain and subtilopeptidase A. Based on known chemistry, the reaction of the enzymes with either activated support (imidocarbonate-containing dextran and diazo-containing starch) is expected to decrease the total number of positively charged groups in the enzymes and to lead to a slight displacement of the pH-activity curve toward alkaline pH values. In the case of S-MDA, a slight shift may also be attributable to the increased ionization of modified tyrosines of the immobilized enzymes. Because of the only slight effect of ionic strength on the pH-activity profiles of these enzymes, both groups of workers suggested that the pH-activity displacements were caused by more localized electrostatic interactions induced presumably by the chemical reactions employed for immobilization.

The pH-activity curve of an immobilized enzyme can be displaced also by an enzymatically generated pH-gradient between the domain of the enzyme-polymer conjugate and the external solution. The magnitude of the displacement is especially dependent on the catalytic activity of the enzyme and the diffusion rate of substrate(s) and product(s) from the site of the reaction. Most examples of enzymatically generated shifts in pH-activity curves have been found with synthetic membrane-enzyme systems (described in Chapter 4), but this behavior has also been observed by Axen and Ernback with enzymes immobilized with agarose gel beads.[48] The pH-activity curves of covalently bound trypsin and chymotrypsin were found to be more narrow and displaced toward higher alkaline pH values relative to the unmodified enzymes. Immobilized chymotrypsin acting on ATEE exhibited a pH optimum of approximately 9.7 compared to 8 for the native enzyme. Likewise, immobilized trypsin acting on *a-N-p*-toluenesulfonyl-L-arginine methyl ester (TAME) exhibited a pH optimum of approximately 9.6 (pH optimum of unmodified enzyme, ca. 7.5). The pH optima of both enzymes, however, were displaced only slightly with casein as the substrate. Papain was also immobilized on cyanogen bromide-activated agarose, but its pH-activity profile with BAEE was essentially equivalent in shape and pH optimum to that of the native enzyme. Axen and Ernback reasoned that this behavior was due to the rapid generation of hydrogen ions within and around the agarose gel beads by the extremely fast conversion of ATEE and TAME by chymotrypsin and trypsin, respectively. These two covalently bound hydrolases hydrolyze the esters to acids and alcohols rapidly, and the hydrogen ions released in the reaction decrease the local pH in and surrounding the gel beads causing a pH gradient to form. Rapid diffusion of the component species of these reactions would be necessary in order to prevent an accumulation of the hydrogen ions and the concomitant drop in pH. The similarity of the pH-activity curve of immobilized papain to the curve of the native enzyme was attributed to the much lower activity of the enzyme with BAEE. Axen and Ernbach also reported visually confirming a lower internal pH of chymotrypsin-agarose conjugates with acid-base indicators and observing a pH optimum of approximately 9 with *physically entrapped* chymotrypsin in non-activated agarose. The latter observation substantiated the results obtained with the covalently bound enzymes. These investigators also pointed out that the large displacements of the pH-activity curves observed for trypsin and chymotrypsin bound to agarose were perhaps not totally attributable only to enzymatically generated pH differences between the domain of the enzyme-polymer conjugate and the external solution. Other factors, chemical modification and charge effects, could also contribute to the overall displacements observed.

In addition to the specific references mentioned

above with regard to a displacement of the pH-activity curve of an immobilized enzyme in either direction from the native enzyme, numerous examples can be found in the literature where no shift occurs at all.[79,104,105,118,122,127,152,154,155,213,216] These include cases in which the charge of the supports is positive, negative or neutral. Presumably, the lack of a shift with charged supports is caused by the operation of two or more of these effects in opposite directions.

**Michaelis Constant**

The determination of the Michaelis constant, $K_m$, of an enzyme for a substrate is one of the most important and useful quantities that can be determined. It is, by definition, the substrate concentration at which v = V/2 in the familiar Michaelis-Menten relationship given in Equation 43.

$$v = \frac{V_{max}}{1 + \frac{K_m}{S}} \qquad (43)$$

In this equation, v is the velocity of the reaction and $V_{max}$ is the maximum velocity obtained when the substrate concentration, S, is high enough to saturate the enzyme. The Michaelis constant is independent of enzyme concentration and is expressed in moles/l. On the other hand, $V_{max}$ is enzyme concentration-dependent.

As alluded to previously, suspended water-insoluble particles are surrounded by an unstirred layer of solvent, and it is worthwhile to discuss briefly the anticipated effect of diffusion on the value of $K_m$. This unstirred layer is known as the Nernst layer, and with water-insoluble enzymes, a concentration gradient of substrate is established across this layer. Consequently, saturation of an enzyme attached to a water-insoluble particle will occur at a higher substrate concentration than normally required for saturation of the freely soluble enzyme. This, in turn, gives a higher $K_m$ value of the immobilized enzyme for its substrate. The Michaelis constant determined for immobilized enzymes is necessarily only an apparent constant ($K'_m$) and should be distinguished clearly from the constant normally determined with the soluble enzyme. Also, if the solvent layer were reduced by methods such as physically reducing the size of the particle which contains the bonded enzyme, or increasing the stirring speed of the suspended particles, or even solubilizing the water-insoluble enzyme derivative, then the $K'_m$ value should be considerably reduced and could even approach the true $K_m$.

Investigations of these considerations have been reported. For example, an increase in the value of $K'_m$ of covalently bonded water-insoluble enzymes for substrates was observed for glutamate dehydrogenase on collagen,[105] cholinesterase on DEAE-cellulose,[96] papain[47] and subtilopeptidase A[47] on S-MDA, β-fructofuranosidase[102] on diazotized polystyrene, chymotrypsin[188] on DEAE-cellulose, glucoamylase[150] on CM-cellulose, and apyrase[140] on CM-cellulose, polygalacturonic acid, and polyaspartic acid. Additional examples of increased $K'_m$ values are available.[64,131,216]

The magnitude of the increase in $K'_m$ is dependent on conditions employed during its determination. In this regard, it is unfortunate that more thorough studies have not been conducted on the examination of this constant over wider ranges of condition variables, for the value obtained at only one state can be misleading. This was shown quite dramatically by Kay and Lilly[188] during the examination of α-chymotrypsin covalently bound to DEAE-cellulose. The variation of $K'_m$ and $K_m$ with pH for the immobilized and the native enzymes, respectively, is quite different over the pH range studied (see Figure 8). Further, even when the reaction rate is partly diffusion-limited, it is possible to obtain data that result in almost straight Lineweaver-Burk plots but that give quite misleading values of $K_m$ and $V_{max}$.

Kay and Lilly[188] also showed that reducing the physical size of a particle containing a covalently bonded enzyme decreases the value of $K'_m$ by reducing the diffusional limitation. Grinding of the α-chymotrypsin-DEAE-cellulose conjugate to a very fine powder reduced the value of $K'_m$ of the enzyme-polymer conjugate for the substrate ATEE. Another interesting study, substantiating the other possibility of reducing the $K'_m$ value of an enzyme attached on a water-insoluble polymer by increasing its solubility in water and thereby largely eliminating diffusional restrictions, was reported by Axen et al.[95] α-Chymotrypsin, covalently bound to water-insoluble Sephadex through the cyanogen bromide-activation procedure, gave in two preparations differing in the pH of the support activation step (pH 10.3 and 9.8) $K'_m$ values of 25 to 30 and 30 to 40 m*M*, respectively, for ATEE. Soluble unmodified α-chymotrypsin exhibited a $K_m$ value of 3.3 m*M*.

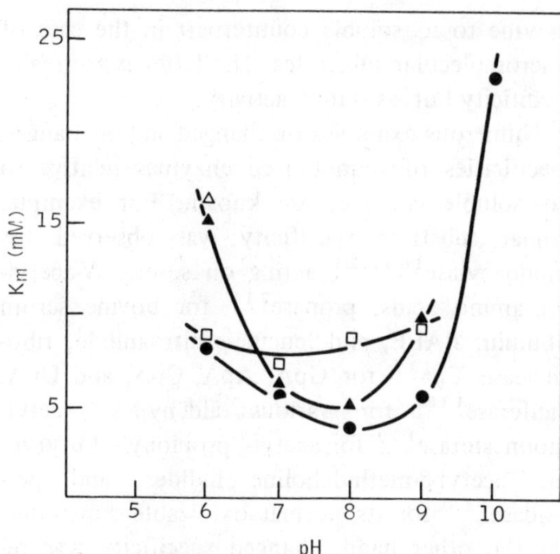


FIGURE 8. Variation of Michaelis constant (apparent and real) with pH for α-chymotrypsin (•) and α-chymotrypsin-DEAE-cellulose (110 mg/g support) before grinding (△) and after grinding (▲). α-Chymotrypsin-DEAE-cellulose (11 mg/g support) (□). Activity measured in presence of dioxane, phosphate buffer, and NaCl. (Redrawn from Kay, G., and Lilly, M. D., *Biochim. Biophys. Acta*, 198, 276, 1970. With permission.[188])

Upon solubilization of these water-insoluble enzyme-polymer conjugates with dextranase, the $K'_m$ values dramatically dropped to a value of 5 and 3 m*M* for the pH 10.3 and 9.8 activated conjugates, respectively. Axen and co-workers also decreased the $K'_m$ of the water-insoluble enzyme by about 25% by grinding the enzyme conjugate to a smaller particle size.

So far, the discussion of the Michaelis constant has been straightforward and has dealt with expected behavior based on diffusion limitations. However, the behavior of the Michaelis constant of many enzyme conjugates is not simple. Many examples are known in which the $K'_m$ of the immobilized enzyme is identical[127,137,157,207,224] to or even smaller than[40,49,63,66,104,116,139,151,211] the $K_m$ of the native, water-soluble enzyme.

The ionic nature of the carrier and substrate can have a profound influence on the magnitude of this constant. Again referring to the studies of Goldstein, Levin, and Katchalski[69] with covalently bonded trypsin on the maleic anhydride ethylene copolymer as the example for illustration, the apparent Michaelis constant of the polyanionic enzyme-polymer conjugate using the positively charged substrate, BAA, was found to be approximately 30 times lower than the native enzyme at low ionic strength. Specifically, the $K'_m$ of the immobilized and soluble trypsin was 0.2 and 6.9 m*M*, respectively, measured at an ionic strength of 0.04 and at their respective optional pH's (9.5 and 7.5). As in the case of the displaced pH-activity curve, increasing the ionic strength of the medium abolished this difference in the magnitude of $K'_m$ (see Figure 9). If the electrostatic interaction between the charged carrier and the charged substrate is responsible for this behavior, it follows that the $K'_m$ of an immobilized enzyme is expected to be lower than the native enzyme whenever the charges are of different polarity, and greater than the native enzyme whenever they are of equal sign. An excellent study supporting this assumption was made by Hornby, Lilly, and Crook,[128] who examined the magnitude of the Michaelis constants of several enzymes attached to differently charged supports and with differently charged substrates. ATP-creatine phosphotransferase bound to CM-cellulose (negatively charged) exhibited a $K'_m$ of 7 m*M* for its substrate ATP (also negatively charged). On the contrary, the $K_m$ of the free enzyme and the $K'_m$

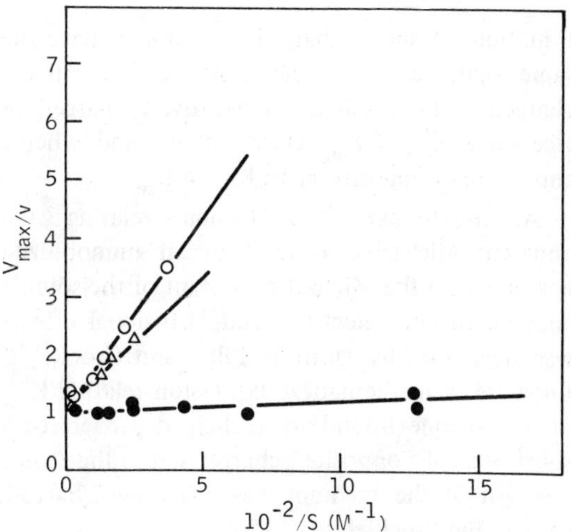


FIGURE 9. Normalized Lineweaver-Burk plots for trypsin and trypsin-maleic anhydride-ethylene copolymer at different ionic strengths with BAA as substrate. Trypsin at an ionic strength of 0.04 (○); $K_m$ 6.8 m*M*. Trypsin-maleic anhydride-ethylene copolymer at ionic strengths of 0.04 (•) and 0.5 (△); $K'_m$ 0.2 and 5.2 m*M*, respectively. (Redrawn from Goldstein, L., Levin, Y., and Katchalski, E., *Biochemistry*, 3, 1913, 1964. With permission.[69])

of the enzyme attached to a neutrally charged carrier, *p*-aminobenzylcellulose, were 0.65 and 0.80 m*M*, respectively. Ficin, on the other hand, attached to CM-cellulose exhibited a $K'_m$ value for BAEE (positively charged) of 2 m*M*; the free enzyme showed a $K_m$ of 20 m*M* for the substrate.

Goldstein, Levin, and Katchalski[69] attributed this phenomenon to an unequal distribution of a charged substrate between the polyelectrolyte phase and the external solution and derived a mathematical expression relating the apparent Michaelis constant of a charged enzyme derivative with the Michaelis constant of the native enzyme to the electrostatic potential (Equation 44).

$$\Delta pK_m = pK'_m - pK_m = \log \frac{K_m}{K'_m} = 0.43 \frac{z\varepsilon\Psi}{kT} \qquad (44)$$

The derivation of Equation 44 followed from the initial relationship between the concentration of substrate in the domain of the enzyme-polymer conjugate, $S^I$ (phase I), and the substrate concentration in the external solution $S^{II}$ (phase II) given in Equation 45. Substituting the value of S given in Equation 45 into the Michaelis-Menten relationship, Equation 44 is obtained.

$$S^I = S^{II} \exp\left(\frac{z\varepsilon\Psi}{kT}\right) \qquad (45)$$

Equation 44 shows that when $z$ and $\Psi$ have the same sign, i.e., when the substrate is positively charged and the carrier is negatively charged or vice versa, $K'_m < K_m$. On the other hand, when $z$ and $\Psi$ are of opposite signs, $K'_m > K_m$.

A mathematical expression relating the apparent Michaelis constant of an immobilized enzyme with the Michaelis constant of the soluble enzyme to both electrical and diffusional effects was developed by Hornby, Lilly, and Crook.[128] Likewise, a mathematical expression relating $K'_m$ of an enzyme bound to a charged carrier for a substrate of opposite charge with the ionic strength of the medium was given by Wharton, Crook, and Brocklehurst.[212]

## Specificity

The term "specificity," as used in this section, has a wider connotation than normally accepted and is used to mean not only the strict stereochemical or electronic requirement needed of a compound to be acted upon by the enzyme but also the relative activity of the immobilized enzyme to its soluble counterpart in the case of macromolecular substrates. The latter is not really specificity but, as stated, activity.

Numerous examples of changed and unchanged specificities of immobilized enzymes relative to the soluble enzymes are known. For example, similar substrate specificity was observed for aminoacylase[164,225] acting on several *N*-acetyl-D,L-amino acids, pronase[216] for bovine serum albumin, BAEE, and leucine-*p*-nitroanilide, ribonuclease $T_1$[137] for GpA, ApA, CpA, and UpA, luciferase[118] for various aldehydes, acetylcholinesterase[139] for acetyl-, propionyl-, butyryl-, and acetyl-β-methylcholine iodides, and peroxidase[125] for its normal oxidizable substrates. On the other hand, changed specificity was reported for immobilized L-amino-acid oxidase,[121] α-amylase[146] and trypsin.[132,166,171-173] Immobilized L-amino acid oxidase, in comparison with the native enzyme, did not act on L-asparagine, and α-amylase exhibited an interesting higher degree of multiple attack on amylose. Various immobilized trypsin derivates present some interesting specificity observations. Trypsin[173] covalently bonded to a maleic anhydride-ethylene copolymer hydrolyzed never more than 10 peptide bonds in pepsinogen while the native enzyme hydrolyzed 15 peptide bonds; different peptide maps were also obtained. Likewise, this water-insoluble trypsin derivative exhibited a lower degree of peptide hydrolysis and showed altered selectivity when employed by Lowey et al.[171,176] for the degradation of myosin and heavy meromyosin. These changed specificities of trypsin immobilized on the charged carriers were attributed to both steric interference caused by the two macromolecular assemblies and by specific charge interactions of the substrate with the carrier.

Inhibition studies have been reported for several immobilized enzymes. Trypsin[45] covalently bonded to AE-cellulose via glutaraldehyde exhibited inhibitory behavior with α-*N*-*p*-toluenesulfonyl-L-lysine chloromethyl ketone and phenylmethanesulfonyl fluoride similar to that with the native enzyme. Surprisingly, the insoluble enzyme could not be completely inhibited by diisopropylphosphorofluoridate and showed a lesser inhibition by arginine. Inhibition of this trypsin derivative by natural protein inhibitors varied from 80 to 100% and depended on the molecular weight of both the inhibitor and the substrate employed for assay. Inhibition of insoluble trypsin by protein inhibitors was reported extensively by Fritz et al.[163,167] Immobilized L-amino acid oxidase[121] exhibited similar substrate inhibition with cysteine, isoleucine, and leucine, as did the native enzyme. β-Fructofuranosidase[102] bound to diazotized polystyrene showed changed inhibitory behavior toward aniline and Tris buffer but still remained noncompetitive in nature; the inhibitor dissociation constant of the insoluble enzyme with respect to Tris increased from 0.45 to 1.10 *M* and with aniline decreased from 0.94 to 0.39 m*M*. Inhibition was also reported for immobilized thrombin,[207] glucoamylase,[150] peroxidase,[125] and ATP-creatine phosphotransferase.[128]

Some specificity studies have also been reported for rather complex enzymes. For example, glutamate dehydrogenase covalently bonded to collagen still retained its regulatory properties when attached to the matrix but lost the linearity in the double reciprocal plot of reaction rate vs. glutamate concentration at all NAD concentrations employed. tRNA nucleotidyltransferase[135] covalently bound to Sepharose exhibited behavior similar to the soluble enzyme concerning the requirement of magnesium ions and mercaptoethanol; however, the modified enzyme had no poly(A)-synthetase activity. Metal specificity studies were reported for ATPase[98,227] and apyrase.[140,230]

Numerous reports have mentioned the difference in the activity of an immobilized enzyme toward high and low molecular weight substrates. In these studies, the comparison of activities is made between the activity of an immobilized enzyme acting on a small-sized substrate and on a large-sized, macromolecular substrate. With the latter compounds, diffusion of the substrate to the active enzyme and steric repulsion of the carrier are expected to necessarily decrease the activity and efficiency of the catalyzed transformation. Indeed, decreased activity was observed for papain[48] bound to Sepharose acting on casein in comparison with BAEE; ribonuclease A[134] attached to agarose acting on RNA in comparison with cytidine cyclic phosphate (CCP); α-chymotrypsin[95] on Sephadex acting on casein in comparison with ATEE; ficin[49] on CM-cellulose acting on casein relative to BAEE; papain[200] on the *p*-aminophenylalanine-L-leucine copolymer acting on casein in comparison with BAEE; and ribonuclease $T_1$[136] bound to CM-cellulose acting on RNA in comparison with CCP.

With regard to the reaction rates of covalently bonded enzymes with different molecular weight macromolecular substrates, Cresswell and Sanderson[213,214] reported an empirical correlation of the molecular weight of a substrate, mol wt, to the soluble enzyme/insoluble enzyme reaction rate ratio, R. Equation 46 applied over a wide size range of substrates.

$$R = \frac{\text{Reaction rate with insoluble enzyme}}{\text{Reaction rate with soluble enzyme}} = K + k \log \text{mol wt} \qquad (46)$$

**Stability**

In discussing the stability of a covalently bonded enzyme-polymer derivative (or any other type of immobilized enzyme), it is essential to state the type of stability one is referring to. That is, is it stability toward heat, extremes of pH, organic solvents, lyophilization, etc.? It is also equally important to give the exact experimental conditions under which the stability of the immobilized enzyme was tested in order to assess whether it indeed has superior stability relative to the water-soluble enzyme.

The stability of an enzyme discussed here and elsewhere usually refers to the conformational stability of the protein. The thermal, organic solvent, urea, pH, and storage stability of an enzyme are largely determined by its conformational stability, and inactivation is caused primarily by denaturation. However, chemical inactivation – the breaking or making of new covalent bonds – could also take place and can contribute to the overall inactivation of the enzyme observed.

*Thermal Stability*

At present, there is considerable confusion in the literature as to whether covalently bonded water-insoluble enzymes *in general* (and other water-insoluble enzyme derivatives as well) have enhanced thermal stabilities. In principle, the thermal stability of an immobilized enzyme can be enhanced, diminished, or unchanged relative to the native, water-soluble enzyme, and numerous examples of each kind exist. Although there are more examples of enhanced thermal stability in the literature, this fact may only be the result of our commonly accepted reporting system; negative results are frowned upon.

Enhanced thermal stabilities have been reported for numerous covalently bonded enzymes, and some recent examples are α-amylase[56,73,75] on Enzacryl AA, AH and Polyacetal, β-glucosidase[152] on cellulose carbonate, amyloglucosidase[151] on DEAE-cellulose, trypsin[164] and aminoacylase[164] on crosslinked acrylamide-methyl acrylate copolymer, invertase[112] on bentonite, tRNA nucleotidyl-transferase[135] on Sepharose, α-chymotrypsin[153] on glass, leucine aminopeptidase[157] on Sephadex, cholinesterase[96] on DEAE-cellulose, a broad specificity protease[221] on Sepharose, luciferase[118] from *Photobacterium fischeri* on polyacrylic acid, L-asparaginase[222] on CM-dextran, CM-cellulose, and cellulose, trypsin[106] on crosslinked acrylamide-acrylic acid copolymer, and urease[224] on nylon. Figures 10 to 12 illustrate the enhanced thermal stabilities observed for several of the enzymes mentioned. The thermal stability of enzymes is usually tested by incubating a suspension of the water-insoluble derivative and the unmodified enzyme at a certain temperature and then measuring the decrease in activity with time.

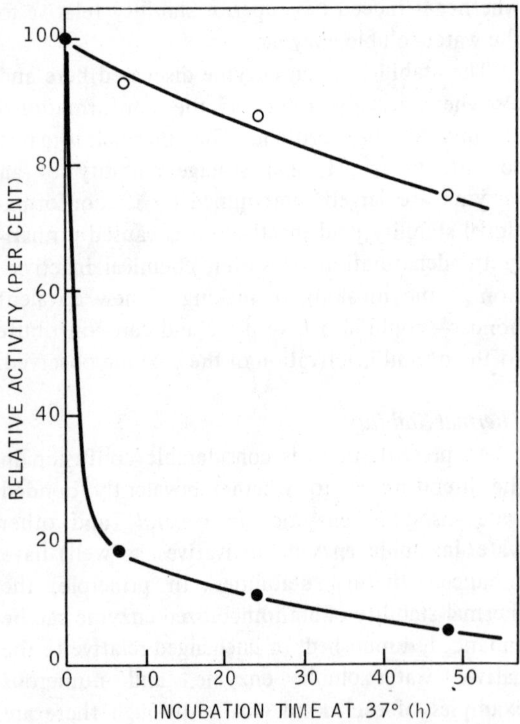


FIGURE 10. Relative activity of β-D-glucosidase (●) and β-D-glucosidase-cellulose carbonate (○) with incubation time at 37° and pH 5.0. (Redrawn from Barker, S.A., Doss, S. H., Gray, C. J., Kennedy, J. F., Stacey, M., and Yeo, T. H., *Carbohyd. Res.*, 20, 1, 1971. With permission.[152])

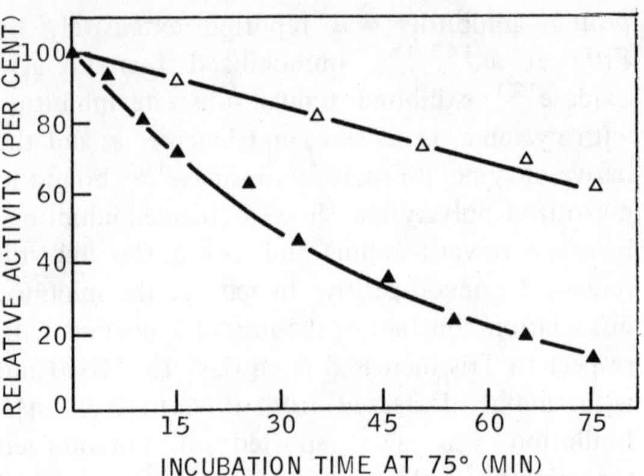


FIGURE 11. Relative activity of urease (▲) and urease-nylon (Δ) with incubation time at 75° and pH 7.0. (Redrawn from Sundaram, P. V. and Hornby, W. E., *FEBS Lett.*, 10, 325, 1970. With permission.[224])

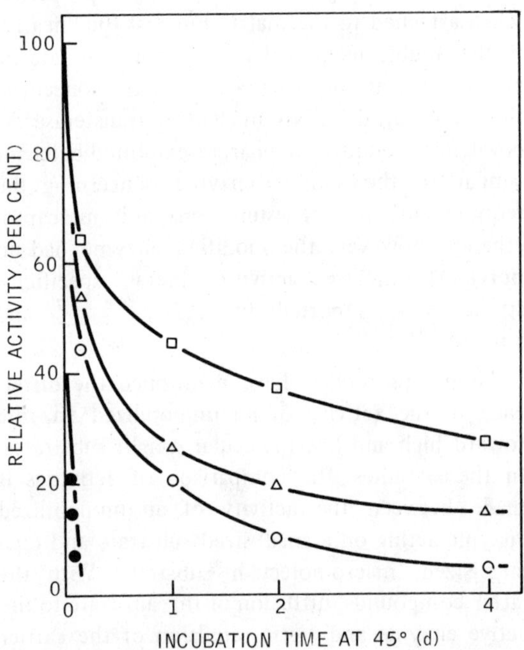


FIGURE 12. Relative activity of α-amylase (●), α-amylase-Enzacryl AA, diazo coupled (○), α-amylase-Enzacryl AA, isothiocyanato coupled (△), and α-amylase-Enzacryl AH (□) with incubation time at 45° and pH 6.9. Redrawn from Barker, S. A., Somers, P. J., Epton, R., and McLaren, J. V., *Carbohyd. Res.*, 14, 287, 1970. With permission.[73])

Water-insoluble enzyme-polymer conjugates of lower thermal stability relative to their water-soluble enzymes are β-amylase[73] covalently bonded to Enzacryl AH via either its isothiocyanato or acylazide derivative (see Figure 13) and luciferase[118] isolated from *Photobacterium*

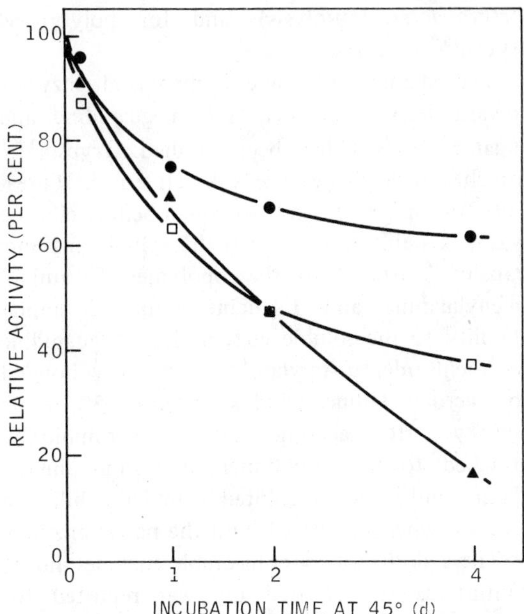


FIGURE 13. Relative activity of β-amylase (●), β-amylase-Enzacryl AA, diazo coupled (▲), and β-amylase-Enzacryl AA, isothiocyanato coupled (□) with incubation time at 45° and pH 4.8. (Redrawn from Barker, S. A., Somers, P. J., Epton, R., and McLaren, J. V., *Carbohyd. Res.*, 14, 287, 1970. With permission.[73])

*leiognathi* bound to polyacrylic acid hydrazide. The decreased thermal stability of the enzyme isolated from *Photobacterium leiognathi* was in direct contrast to the case of the enzyme bound on the same carrier but isolated from *P. fischeri.* Decreased thermal stability was also reported for L-asparaginase[222] bound to CM-cellulose via Woodward's Reagent K, glucoamylase[150] on CM-cellulose, peroxidase[125] on CM-cellulose, papain[200] on the copolymer of *p*-aminophenylalanine and L-leucine, and alcohol dehydrogenase[110,111] on a nitrated copolymer of methacrylic acid and methacrylic acid-*m*-fluoroanilide. Decreased temperature optima of immobilized enzymes were found for pronase[213] covalently bound to the copolymer of *p*-aminophenylalanine and L-leucine, trypsin[123] on cellulose, and *crude* pepsin[161] on glass. On the contrary, immobilized crystalline pepsin had a higher temperature optimum than the native enzyme.

Similar thermal stabilities of the immobilized enzyme relative to the native enzyme were observed for cholinesterase[139] attached to Sepharose and apyrase[230] bound to CM-cellulose.

Examples of increased and decreased thermal stabilities of covalently bonded, *water-soluble* enzyme polymer conjugates are also known.[90,91]

*Storage Stability*

Water-insoluble enzyme polymer conjugates are often stored in appropriate buffers at or near 5° or as lyophilized preparations if this operation does not inactivate the bound enzyme. Lyophilization of covalently bonded enzyme-polymer conjugates was reported for several enzymes[156,218] but in most cases this led to considerable inactivation.[47,64,147,200] Examples of long-term stability measurements (with the relative activity remaining at the time indicated) were reported for β-galactosidase[97] on DEAE-cellulose (19% after 3 years at 2 to 5°), ribonuclease $T_1$[137] on Sepharose (87% after 35 days at 4°), ribonuclease A[134] on agarose (100% after 5 months at 4°), α-amylase[75] on Enzacryl AA and AH (73 and 67% for the diazo and isothiocyanato AA derivatives, respectively, after 3 months at 0 to 5° and 85% for the AH derivative), α-amylase[147] on cellulose (62 to 71% after 4 months at 0 to 5°), α-chymotrypsin on cellulose (60% after 2 to 2.5 years at 2°), and papain[200] on the copolymer of *p*-aminophenylalanine and L-leucine (25% after 9 months at 4°). Additional references to storage stabilities of immobilized enzymes have been published.[64,65,73,86,92,102,141,156,161] Decreased storage stability of an immobilized enzyme was also reported; glucose oxidase[113] covalently bound to CM-cellulose retained less activity than the soluble native enzyme at 5°.

*pH Stability*

Only a limited number of studies have concerned themselves with the pH stability of covalently bonded, water-insoluble enzymes. Polytyrosyl trypsin[47] attached to S-MDA exhibited superior stability at alkaline pH's compared with the native or polytyrosyl trypsin (see Figure 14). On the other hand, papain and subtilopeptidase A bound to S-MDA showed the same and an enhanced (toward lower pH's) stability, respectively Improved stabilities toward alkaline pH's were reported for polytyrosyl trypsin[55] attached to the copolymer of *p*-aminophenylalanine and L-leucine. pH-stability studies of covalently bound enzymes were reported for naringinase[156] and glucoamylase.[150]

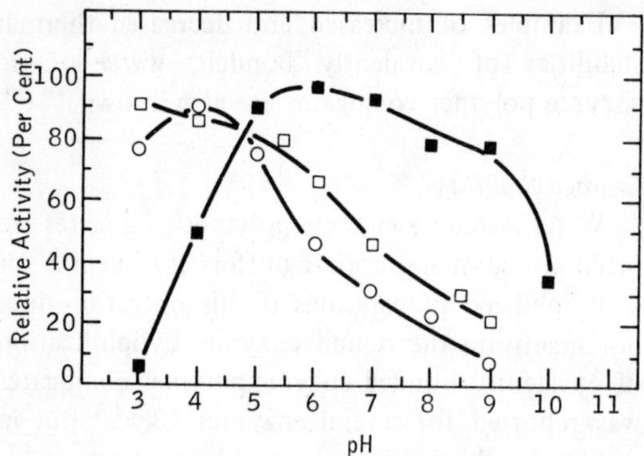


FIGURE 14. pH-Stability of trypsin (○), polytyrosyl trypsin (□) and polytyrosyl trypsin-S-MDA (■). Relative activity of trypsin and trypsin derivatives determined with BAEE at normal assay pH's (8 and 10 for trypsin and polytyrosyl derivatives, respectively) after incubation for 30 min at 37° at the indicated pH. (Redrawn from Goldstein, L., Pecht, M., Blumberg, S., Atlas, D., and Levin, Y., *Biochemistry,* 9, 2322, 1970. With permission.[47])

*Miscellaneous Stabilities*

Enhanced stability toward proteolysis of immobilized enzymes was observed for tRNA nucleotidyltransferase[135] (against trypsin degradation), for the protease[221] isolated from an *Arthrobacter* (autolysis), and for polytyrosyl trypsin[52] (autolysis).

The stability of some immobilized enzymes toward denaturants such as urea, guanidine, and organic solvents has been studied. Trypsin[183] attached to Sephadex was still active in 8 *M* urea, but trypsin[123] bound to aminobenzylcellulose was less stable in urea than the soluble enzyme. Papain[200] attached to the copolymer of *p*-aminophenylalanine and L-leucine exhibited similar stability to the soluble enzyme in 4 *M* guanidine hydrochloride. α-Amylase[147] covalently bonded to microcrystalline cellulose retained 5% of its activity after acetone drying. Thrombin[206] attached to the copolymer of *p*-aminophenylalanine and leucine exhibited a similar stability in glycerol when compared with the native enzyme.

The stability of water-insoluble enzymes during continuous or repeated use was reported for trypsin[76] and papain[76] on Enzacryl Polyacetal, tyrosinase[122] and penicillin amidase[66] on DEAE-cellulose, glutamate dehydrogenase[105] on collagen, and trypsin[101] on nylon.

Enhanced resistance toward oxidation was reported for bromelain[211] and for ficin[131] attached to CM-cellulose.

Chapter 4

# INTERMOLECULAR CROSSLINKING USING MULTIFUNCTIONAL REAGENTS

The preparation of water-insoluble enzyme derivatives using low molecular weight multifunctional reagents involves the covalent bond formation between molecules of the enzyme and the reagent to give intermolecularly crosslinked species. The reactions are irreversible, but the irreversibility is not an inherent characteristic of the method. It only reflects the fact that the chemical reactions used for preparing these immobilized enzyme derivatives have been of such a nature.

A number of distinctly different ways are available for preparing water-insoluble enzyme derivatives using low molecular weight multifunctional reagents. As discussed in Chapter 2, multifunctional reagents can be used to introduce functional groups into preformed, water-insoluble polymers which then react covalently with water-soluble enzymes. Two prevalent examples of this method are the glutaraldehyde modification of acrylamide and the reaction of the various *s*-triazinyl derivatives with celluloses. In this method, the multifunctional reagent is employed only as a *monofunctional* reagent in its reaction with the enzyme. One part of the reagent is used to introduce an appropriate functionality into the preformed polymer and to serve as a means for extending the reactive functional group from the polymer's backbone, and the other part is used to bind the enzyme. Other methods also exist, but before discussing these, it is instructive to bring out some general considerations of multifunctional reagents.

A large number of different multifunctional reagents have been employed to measure inter-residue distances in proteins, to bind two different macromolecules to each other, to modify cell membranes or other multimacromolecular assemblies, and to introduce additional tertiary structure into enzymes with the hope of increasing their conformational stability. Excellent reviews on the utility and scope of multifunctional reagents in protein chemistry are those of Wold[235] and Fasold et al.[236] In the above-mentioned uses of multifunctional reagents, the experimental conditions were chosen so that only a limited modification of the enzymes was obtained and no water-insoluble derivatives were produced. The correct choice of reaction conditions is indeed one of the major considerations in dealing with the chemical modification of proteins with multifunctional reagents for producing *water-insoluble* enzyme derivatives. The concentration of the enzyme and the reagent, the pH and ionic strength of the solution, etc. must be chosen carefully so as to be favorable for *intermolecular* crosslinking between enzyme molecules. Depending on the conditions, the molecular weight and solubility of modified enzymes can range from highly soluble monomeric species to oligomeric complexes which are still water-soluble, to three-dimensional, crosslinked species that are completely insoluble in water.

A second important consideration is the degree of *intramolecular* crosslinking. *Intramolecular* crosslinking of an enzyme with a multifunctional reagent is favored by a low concentration of the enzyme and the reagent. Low concentrations enhance the probability of the functional groups of a reagent molecule to react with amino acid residues only of the same enzyme molecule and not of different molecules. The control of *intra*- vs. *intermolecular* crosslinking is, however, experimentally very difficult to achieve and immobilization of an enzyme with multifunctional reagents will necessarily involve both types of crosslinks. Exact conditions chosen for insolubilization will influence also the degree of intramolecular crosslinking. Furthermore, the reaction of a multifunctional reagent with an enzyme need not involve only the creation of *intra*- or *intermolecular* crosslinks. It can involve only one functional group of the multifunctional reagent; the other end of the molecule can be either left as such or transformed to a nonreactive group through the action of the solvent.

Multifunctional reagents that have been employed to date for the preparation of water-insoluble enzyme derivatives are given in Figure 15. The reagents are all bifunctional in nature and have the same organic functionality for each reactive center. Surprisingly, hetero bifunctional reagents have not yet been used for this purpose. Most of the reagents are available commercially or can be synthesized easily. Several of these are also soluble in water. The structure and chemistry of the reagents, especially glutaraldehyde, are discussed later.

In this chapter are described also water-insoluble enzyme derivatives prepared by highly reactive low molecular weight reagents that cause a condensation among individual protein molecules. The chemistry of these reagents is completely different from those shown in Figure 15. These condensation reagents are not incorporated into the structure of the water-insoluble enzyme derivative; figuratively speaking, they cause the elements of water to be removed from two or more enzyme molecules.

## Methods for Preparing Water-Insoluble Enzyme Derivatives

A general description of the different methods for the preparation of water-insoluble enzyme derivatives with low molecular weight multifunctional reagents is given in outline form in Table 6.

TABLE 6

**Variations for Producing Water-Insoluble Enzyme Derivatives with Multifunctional Reagents**

Reaction of the enzyme with the:

A. Multifunctional reagent alone
B. Multifunctional reagent in the presence of a second protein (coprotein)
C. Multifunctional reagent after adsorption of the enzyme on a water-insoluble, surface-active support
D. Functionalized water-insoluble support prepared previously by the reaction of the multifunctional reagent with a preformed, water-insoluble polymer

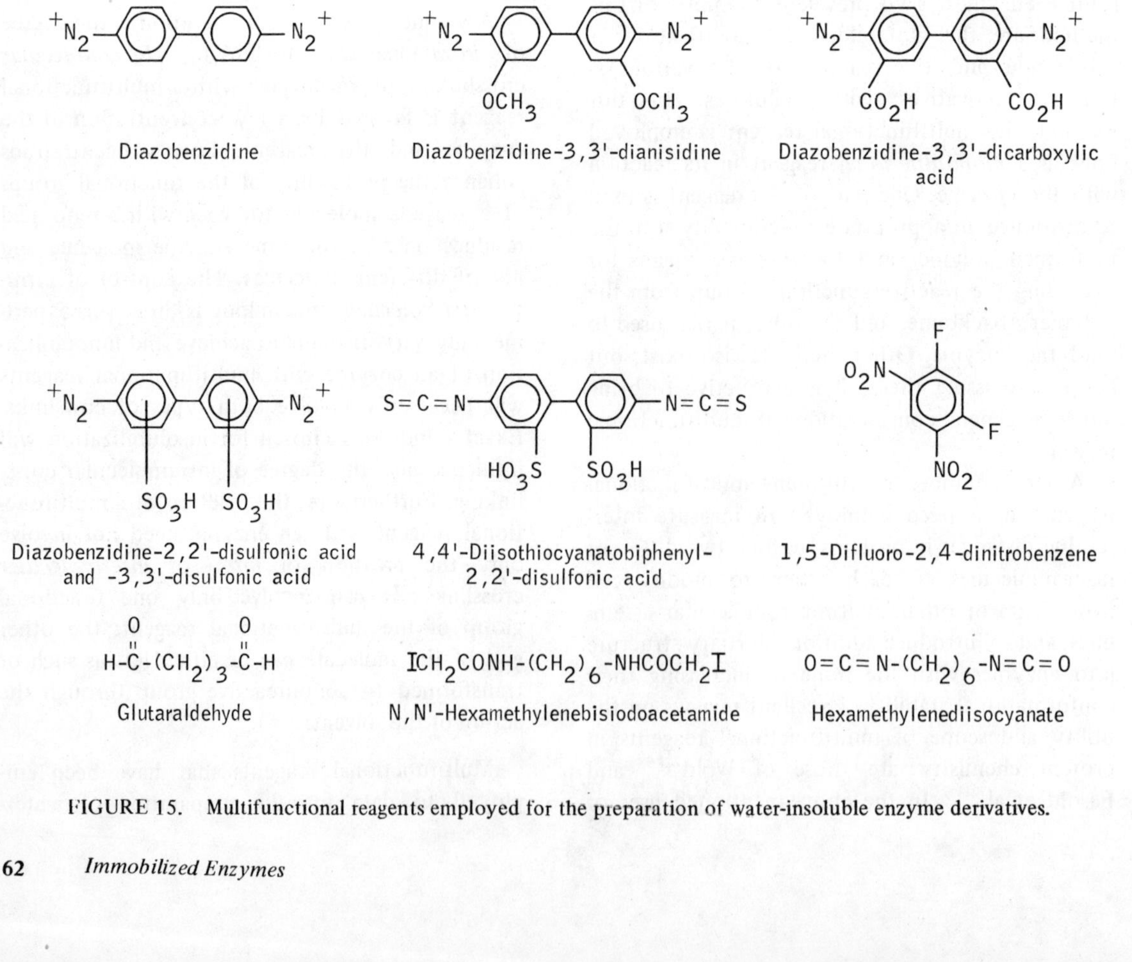

Diazobenzidine

Diazobenzidine-3,3'-dianisidine

Diazobenzidine-3,3'-dicarboxylic acid

Diazobenzidine-2,2'-disulfonic acid and -3,3'-disulfonic acid

4,4'-Diisothiocyanatobiphenyl-2,2'-disulfonic acid

1,5-Difluoro-2,4-dinitrobenzene

$H\text{-}C(=O)\text{-}(CH_2)_3\text{-}C(=O)\text{-}H$

Glutaraldehyde

$ICH_2CONH\text{-}(CH_2)_6\text{-}NHCOCH_2I$

N,N'-Hexamethylenebisiodoacetamide

$O{=}C{=}N\text{-}(CH_2)_6\text{-}N{=}C{=}O$

Hexamethylenediisocyanate

FIGURE 15. Multifunctional reagents employed for the preparation of water-insoluble enzyme derivatives.

Only the first three methods are discussed in this chapter; method D is covered in Chapter 2 and examples of this type of immobilization are given in Table 4.

The most commonly employed method for preparing water-insoluble enzyme derivatives is through intermolecular crosslinking of enzyme molecules with multifunctional reagents. Method A involves the addition of an appropriate amount of bifunctional reagent to an enzyme solution at conditions that give the desired water-insoluble derivative. Although this procedure is operationally simple, it is far from being that simple experimentally. Optimum conditions for achieving good insolubilization while retaining considerable enzymic activity must be determined by trial and error. Only a small number of relatively thorough studies have been reported on the preparation of water-insoluble enzyme derivatives by this technique,[237-240] but these have shown that insolubilization is critically dependent on a delicate balance of factors such as the concentration of the enzyme and multifunctional reagent, the pH and ionic strength of the solution, the temperature, and the time of the reaction. Certain results from these studies illustrate the delicate balance quite clearly.

Jansen and Olson[237] observed that the insolubilization of papain with glutaraldehyde is pH-dependent; the higher the pH, the more rapid the reaction. Complete insolubilization occurred in 24 hr in the pH range of 5.2 to 7.2 at 0° and at a glutaraldehyde and protein concentration of 2.3 and 0.19%, respectively. Yet no precipitation occurred in 24 hr at pH 4.6, even though a reaction between glutaraldehyde and papain was occurring (evidenced by the loss of esterase activity). Jansen, Tomimatsu, and Olson[238] reported further on the extent of insolubilization of proteins as a function of pH. The pH values for the most rapid insolubilization of bovine serum albumin, soybean trypsin inhibitor, lysozyme, and papain were found to be nearly the same as the isoelectric points of these proteins, but $\alpha$-chymotrypsin and $\alpha$-chymotrypsinogen A differed. The formation of insoluble and active $\alpha$-chymotrypsin was most rapid at pH 6.2 (isoelectric point, pH 8.6) and $\alpha$-chymotrypsinogen A at pH 8.2 (isoelectric point, pH 9.5). With $\alpha$-chymotrypsin, at pH 6.4 only 70% of the protein precipitated, and at pH 6.8, no precipitation occurred at all. Likewise, at pH values 6.0 and 5.9, 68 and 56% of the protein precipitated, respectively. At pH 5.7 and below, no insolubilization occurred. The rate of inactivation of $\alpha$-chymotrypsin also increased with increasing pH, and the degree of insolubilization was greater at lower ionic strengths. In a subsequent study by Tomimatsu et al.,[239] it was shown by light scattering measurements that the formation of the insoluble, enzymatically active derivative occurred via a two-step reaction, and that the effects of pH and ionic strength on the rate of the second step of the reaction (the second step being a condensation polymerization reaction) were best explained in terms of an acid shift in the $pK_a$ of the $\epsilon$-amino groups of glutaraldehyde-modified lysine residues and a decrease in attractive forces between enzyme particles with increasing ionic strength. There was also a great loss in enzymic activity (60 to 70%) during the formation of relatively small, but still water-soluble, enzyme derivatives, indicating that the major reason for the reduced activity of water-insoluble enzyme derivatives was chemical modification and not steric or diffusional difficulties.

Ottesen and Svensson[240] reported a dramatic example of the sensitivity of insolubilization to temperature. Extensively substituted, but still water-soluble, derivatives of papain were produced by reaction with glutaraldehyde at 0° (pH 6.0) and at low concentrations. These derivatives were more than 80% modified with glutaraldehyde and remained water-soluble even after the solution was brought to room temperature. In contrast, when the same reaction was conducted at room temperature and pH 6.0, even a tenfold lower glutaraldehyde concentration was sufficient to cause the formation of an insoluble derivative within half an hour. It is interesting that these investigators observed opposite behavior with soluble papain derivatives to that reported for $\alpha$-chymotrypsin by Tomimatsu et al.[239] All the soluble papain derivatives retained about 90% of the initial proteolytic activity even at 80% modification of all the available amino groups. Ottesen and Svensson reasoned that in the insoluble glutaraldehyde-papain derivatives, the amino groups could only be substituted to a smaller degree before the activity drastically decreased, and suggested that the major factor responsible for the decrease in the specific activity of the insoluble enzyme derivatives was the formation of the three-dimensional polymeric protein network.

Water-insolubilization of enzymes can be achieved by treatment of whole enzyme crystals with solutions of multifunctional reagents. This was done with several enzymes primarily for the purpose of enhancing the mechanical strength of crystals in order that they could be more easily handled in x-ray diffraction studies. Quiocho and Richards[241-243] treated carboxypeptidase A crystals with glutaraldehyde and showed that they still retained substantial activity and that the crosslinking occurred throughout the entire crystal and not just at the surface. These crosslinked crystals were used also for metal-exchange studies. Other enzyme crystals crosslinked with multifunctional reagents are ribonuclease A with 1,5-difluoro-2,4-dinitrobenzene[244] and lysozyme with glutaraldehyde.[245]

The preparation of water-insoluble enzyme derivatives with multifunctional reagents can be achieved even after entrapment of the enzyme in either nylon or collodion microcapsules[246] (see Chapter 7). Catalase was first immobilized by Chang, using his normal microencapsulation procedure and then water-insolubilized by treatment of the microcapsules with glutaraldehyde.

The known examples of the preparation of water-insoluble enzyme derivatives by the reaction of multifunctional reagents with only the enzyme, as well as the other methods discussed here, are found in Table 7.

The preparation of water-insoluble enzyme derivatives with multifunctional reagents in the presence of a second protein (method B) is advantageously employed whenever extensive modification of the enzyme leads to gross inactivation or when its supply is limited. Operationally, the same considerations apply as to the previously discussed method for cross linking pure enzymes. Proteins that have been used for this purpose are bovine serum albumin, human immunoglobulin-G, human serum albumin, and sheep and rabbit antibody.[247-250] In some examples,[248] only water-soluble derivatives were reported. Enhanced insolubilization of enzymes with bifunctional reagents was reported for trypsin[166] with glutaraldehyde in the presence of ammonium sulfate, and similarly, for subtilopeptidase B[251] (subtilisin Nova) with ammonium sulfate or acetone.

Water-insoluble enzyme derivatives can be prepared by adsorption of the protein on surface-active water-insoluble supports followed by intermolecular crosslinking with multifunctional reagents. Supports that have been used to produce *surface-layered* water-insoluble enzyme derivatives are colloidal silica[159,252-256] and Dow-Corning 500-3 Silastic® sheets.[257] Intermolecularly crosslinked *impregnated* water-insoluble enzyme derivatives were prepared from porous collodion membranes[258-261] and cellophane sheets.[250,262-265] Goldman et al.[259-261] investigated extensively the preparation and behavior of enzyme-collodion membranes and reported the preparation of a pure papain membrane[259] that was ingeniously produced by simply dissolving the collodion of a previously synthesized papain-collodion membrane with methanol.

**Chemistry of Multifunctional Reagents with Enzymes**

The chemical reactions of most multifunctional reagents shown in Figure 15 are discussed in Chapter 2. Diazo compounds can react with lysine, histidine, tyrosine, arginine, or cysteine residues of proteins. Isocyanates react predominantly with primary amino groups to form substituted ureas and thioureas, respectively. Activated aryl fluorides and alkyl iodides likewise react with nucleophilic groups; aryl fluorides react with lysine residues and iodoacetamides combine with cysteine residues. Carbonyl-containing compounds react predominantly with lysine residues. Not shown in Figure 15 are the condensation reagents that have been employed for preparing water-insoluble enzyme derivatives. Only two such reagents have been used to date for this purpose; ethyl chloroformate[249,266,267] and *N*-ethyl-5-phenylisoxazolium-3′-sulfonate (Woodward's Reagent K),[268] whose structures are shown below. The proposed pathways for the condensation reactions of these reagents with enzyme molecules are given in Equations 47 and 48.

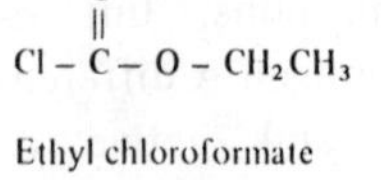

$Cl-C(=O)-O-CH_2CH_3$

Ethyl chloroformate

$SO_3^-$ ... $N^+-CH_2CH_3$

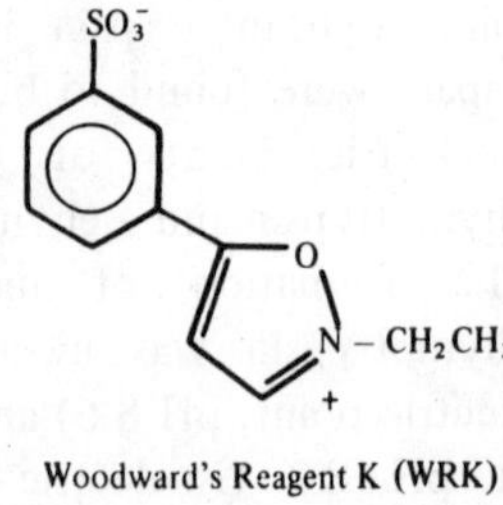

Woodward's Reagent K (WRK)

In direct contrast to the behavior of the multifunctional reagents shown in Figure 15, these two reagents, in principle, are not incorporated into

$$\mathrm{E-CO_2^- + Cl-\overset{\overset{\large O}{\|}}{C}-OEt \longrightarrow E-\overset{\overset{\large O}{\|}}{C}-O-\overset{\overset{\large O}{\|}}{C}-OEt \xrightarrow{E-NH_2} E-\overset{\overset{\large O}{\|}}{C}-NH-E \rightarrow\rightarrow Water\text{-}insoluble\ enzyme} \tag{47}$$

$$\mathrm{+\ CO_2 + EtOH}$$

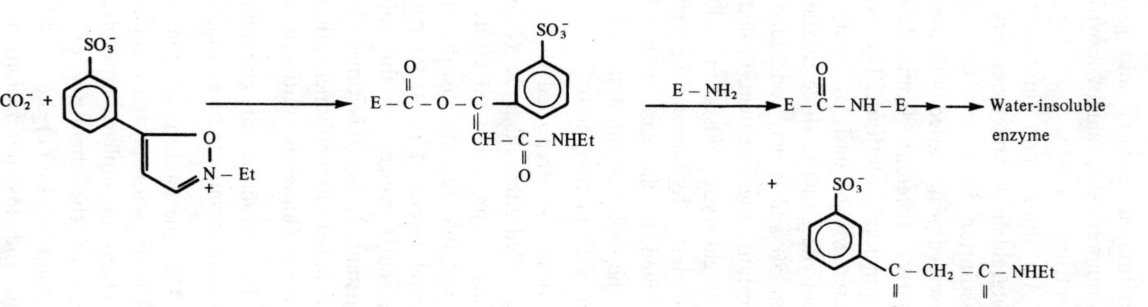

(48)

$$\mathrm{H-\overset{\overset{\large O}{\|}}{C}-CH_2-CH_2-CH_2-\overset{\overset{\large O}{\|}}{C}-H \longrightarrow H-\overset{\overset{\large O}{\|}}{C}-CH_2-CH_2-CH_2-CH=\overset{\overset{\large CHO}{|}}{C}-CH_2-CH_2-\overset{\overset{\large O}{\|}}{C}-H \longrightarrow}$$

$$\mathrm{H-\overset{\overset{\large O}{\|}}{C}-CH_2-CH_2-CH_2-CH=\overset{\overset{\large CHO}{|}}{C}-CH_2-\overset{\overset{\large CHO}{|}}{C}=CH-CH_2-CH_2-CH_2-\overset{\overset{\large O}{\|}}{C}-H \longrightarrow} \tag{49}$$

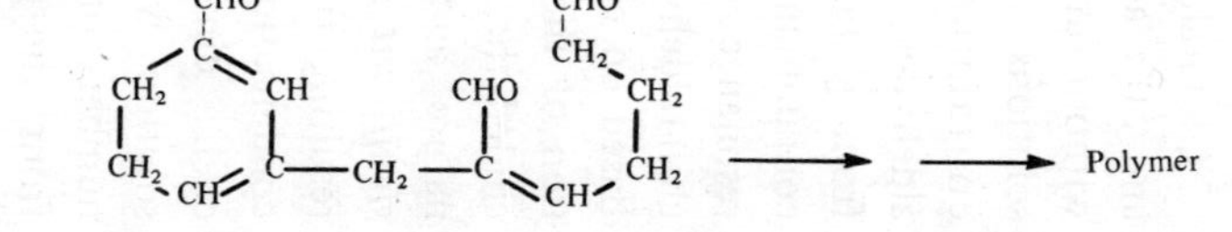

$\longrightarrow \longrightarrow$ Polymer

the final water-insoluble enzyme derivative.

The bifunctional aldehyde, glutaraldehyde, is the most prevalent multifunctional reagent used for the insolubilization of enzymes (see Table 7). It is quite surprising to discover, therefore, that the chemistry and even the structure of this seemingly simple reagent are far from being understood. Because this reagent is so frequently used in protein chemistry, its structure often simply taken as that given in Figure 15, and its chemistry incorrectly assumed to be straightforward, the structure and reactions of this important bifunctional reagent are discussed below.

*Structure and Chemistry of Glutaraldehyde*

Glutaraldehyde, available commercially in 25 and 50% aqueous solutions, is used most often without further purification. Unfortunately, these solutions are invariably not pure and can be contaminated with acids, *a,β*-unsaturated aldehydes, and polymeric materials originating from glutaraldehyde itself. Evidence for these contaminants came from a proton magnetic resonance (pmr) study of commercial aqueous glutaraldehyde by Richards and Knowles.[269] Based on chemical shifts and the relative integration of the protons, Richards and Knowles concluded that glutaraldehyde is largely polymeric in nature and that commercial solutions contain *significant* amounts of *a,β*-unsaturated aldehydes resulting from the loss of water through aldol condensation reactions. The possibility of linear or cyclic dimers or trimers of glutaraldehyde in solution was discounted on the basis of the numbers of different types of protons observed from the spectrum integration. The reaction sequence of glutaraldehyde leading to the formation of the polymeric material via repeated aldol condensations, and its probable structure, as suggested by Richards and Knowles, are given in Equation 49.

Hardy, Nicholls and Rydon[270] similarly examined aqueous glutaraldehyde solutions by pmr spectroscopy, but they proposed a different structure for the dialdehyde. Although two commercial 25% solutions of glutaraldehyde showed peaks in approximately the same positions as those observed by Richards and Knowles, the relative areas of these peaks differed in the two samples and from those found earlier. Hardy et al. also observed a weak 235 nm ultraviolet absorption band in the samples ($E_{1\,cm}^{1\%}$ = 1.5 and 6.2) and attributed this to the presence of unsaturated compound(s) which were most likely *a,β*-unsaturated aldehyde(s). However, the low intensity of the bands suggested that these *a,β*-unsaturated aldehyde(s) were *only minor component(s)* of the total organic material in the solutions. Further, saturation of the commercial solutions with sodium chloride, extraction with ether, and subsequent distillation eliminated the unsaturated material and gave a 50% yield of monomeric glutaraldehyde with the expected pmr spectrum.

Additional pmr investigations showed that glutaraldehyde undergoes very rapid hydration on dissolution in water; the relative areas of the observed peaks varied with concentration and were clearly different from those observed with commercial solutions. These results were interpreted on the basis of equilibria between glutaraldehyde and the three hydrates shown in Equation 50 and it was concluded (from equilibrium constant measurements) that very little free glutaraldehyde exists in solution. The almost complete hydration of glutaraldehyde was ascribed to the ease with which the two ends of the molecule can approach each other to form the cyclic monohydrate.

It was also reported that aqueous solutions of redistilled glutaraldehyde exhibited essentially the same pmr and ultraviolet absorption (no appreciable 235 nm absorption) spectra even after several weeks at room temperature and that the molecular weight of the solute also remained constant during this time. On this basis, it was concluded that polymerization of glutaraldehyde in *neutral* aqueous solution is very slow.

The structure of glutaraldehyde in aqueous solution remains to be established.

The mechanism of the reaction of glutaraldehyde with proteins and the nature of the products are similarly unresolved. Major characteristics of the chemistry of glutaraldehyde with proteins[269] are (1) the overall reaction (modification and insolubilization) is rapid and usually complete within an hour under normal circumstances; (2) the reaction can be considered a two-step process;[239] the first step being the formation of soluble intermolecularly crosslinked complexes followed by a second rapid step leading to insolubilization; (3) insolubilization is de-

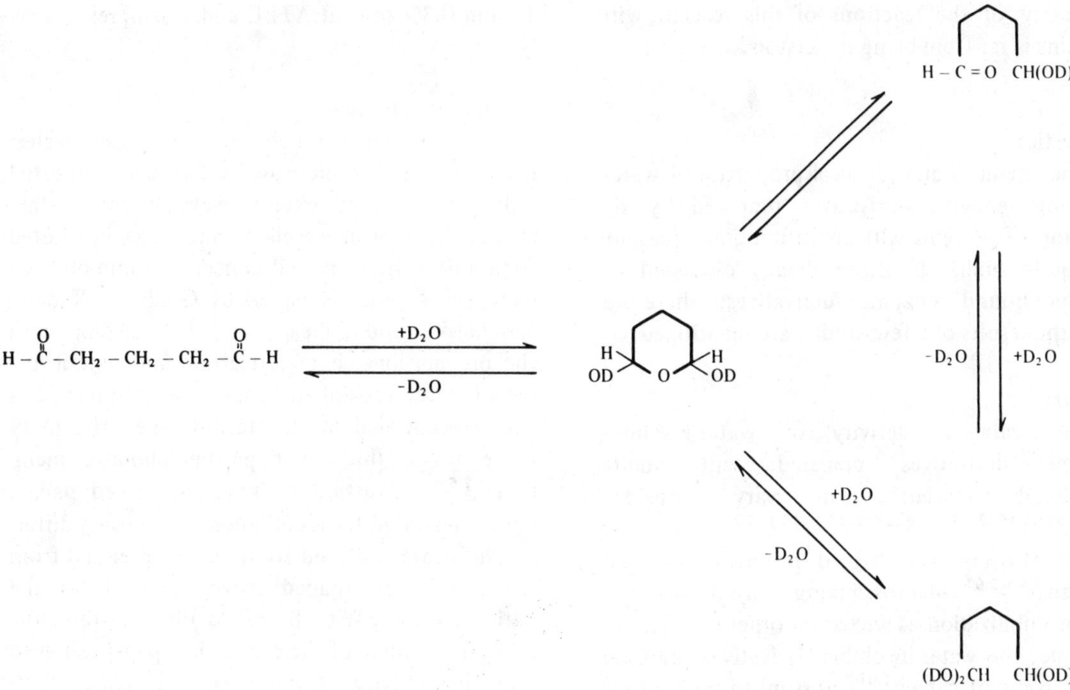

$$H-\overset{O}{\overset{\|}{C}}-CH_2-CH_2-CH_2-\overset{O}{\overset{\|}{C}}-H \underset{-D_2O}{\overset{+D_2O}{\rightleftharpoons}}$$

H, OD, O, H, OD (cyclic hydrate)

H – C = O  CH(OD)$_2$

$-D_2O$  $+D_2O$

$+D_2O$  $-D_2O$

(DO)$_2$CH  CH(OD)$_2$

(50)

CHO  CHO  CH  CH$_2$  CH  $+ E-NH_2 \longrightarrow$  CHO  CHO  CH  CH  CH  CH$_2$  CH  NH  NH  E  E

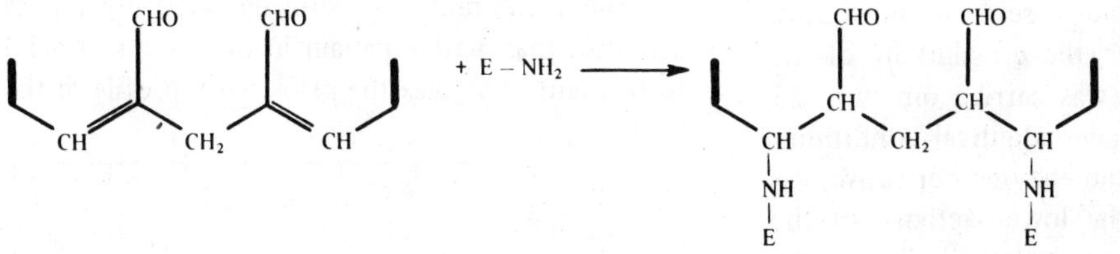

(51)

pendent on the pH, temperature, and ionic strength of the solution; (4) protein modification (after insolubilization) is apparently irreversible; (5) amino acid analyses of acid hydrolysates of several modified proteins reveal consistently a significant loss in the total number of lysines although tyrosine, histidine, and cysteine residues have also been implicated;[240,270a] (6) a new ninhydrin positive material can be seen in standard amino acid chromatograms of treated protein hydrolysates (at least in the case of papain[240] treated with glutaraldehyde in the presence of sodium borohydride); and (7) the observed acid shift in the $pK_a$ of the glutaraldehyde-modified proteins (a decrease in the apparent $pK_a$ of 0.5 to 1 unit was noted for β-lactoglobulin near pH 9).[269]

Based on their pmr study, which indicated that glutaraldehyde exists largely as a polymeric material, Richards and Knowles also proposed a mechanism for the reaction of this bifunctional reagent with proteins. The essential feature of their mechanism is the Michael-type addition of amino groups of the protein to the α,β-unsaturated aldehydic polymer (see Equation 51).

Hardy et al., based on the results of their pmr study of glutaraldehyde, differed again with Richards and Knowles on the mechanism of reaction of this reagent with proteins. Because polymerization of glutaraldehyde in neutral solution was found to be very slow, they suggested that crosslinking under these conditions involved glutaraldehyde itself or one of its hydrates and not an unsaturated oligomer or polymer. They did, however, add that glutaraldehyde polymerizes rapidly in weakly alkaline solutions (pH 8) to an insoluble solid whose spectroscopic properties are consistent with the unsaturated structures proposed by Richards and Knowles, and that crosslinking in alkaline solutions may indeed involve such species.

As with the structure of glutaraldehyde, the

chemistry of the reactions of this reagent with proteins is far from being understood.

**Properties**

The chemical and physical properties of water-insoluble enzyme derivatives prepared by the reaction of proteins with multifunctional reagents are quite similar to those already discussed for polymer-bound enzyme derivatives; therefore, only the results of a few studies are mentioned.

*Activity*

The catalytic activity of water-insoluble enzyme derivatives prepared with multi-functional reagents can vary considerably[166,200,237,238,240,251,271,272] and has been shown to be dependent on the amount[200,268] of crosslinking reagent used during insolubilization as well as on other factors. For example, the water-insoluble derivatives prepared from 56 mg of papain[200] crosslinked with 15 mg of diazobenzidine exhibited a 19 and 6% relative activity with BAEE and casein as substrates, respectively. If, however, the crosslinking of the same amount of papain was carried out with 25 mg of diazobenzidine under identical conditions, the esterase activity of the enzyme derivative was only 8%. Presumably, the lower activity of the second preparation is due to either a higher degree of modification of the enzyme molecules or to diffusional effects.

Considerable enzymic activity was even observed for carboxypeptidase A crystals[241,242] crosslinked with glutaraldehyde. The enzymic activity of such crystals for the substrate, carbobenzoxyglycyl-L-phenylalanine, varied from 30 to 70% relative to unmodified crystals. Unfortunately, the enzymatic activity of other crosslinked enzyme crystals, ribonuclease[244] and lysozyme,[245] was not reported.

Jansen, Tomimatsu and Olson[238] described the activity of a "double enzyme" prepared by the reaction of glutaraldehyde with α-chymotrypsin and mercuripapain. After activation of mercuripapain with cysteine, α-chymotrypsin exhibited a relative activity of 12 and 0.1% toward ATEE and casein, respectively, and papain exhibited a relative activity of 65 and 8% toward BAEE and casein, respectively. In comparison, α-chymotrypsin crosslinked with glutaraldehyde in the presence of bovine serum albumin showed a relative activity of 12 and 0.3% toward ATEE and casein, respectively.

*pH-Activity Behavior*

The pH-activity behavior of these water-insoluble enzyme derivatives has been reported only to a limited extent. Nevertheless, certain studies have been excellent and have illustrated some rather fundamental concepts of immobilized enzymes. A series of papers by Goldman, Kedem, Katchalski, and colleagues[258-261] dealing with the preparation, characterization, and kinetic behavior of crosslinked enzyme-membranes has been exceptional in this regard. The pH-activity curve for a three-layer papain-collodion membrane,[259] consisting of two crosslinked papain layers separated by a collodion layer, using different substrates differed from one another and from the normal bell-shaped curve obtained for the native enzyme. With BAEE as the substrate, the pH-activity curve of the crosslinked papain showed that the activity of the enzyme increased with increasing pH even up to a value of 9.6. The enzyme membrane also showed relatively higher activities than native papain in the region of pH 3 to 6. Figure 16 gives the pH-activity profile of the

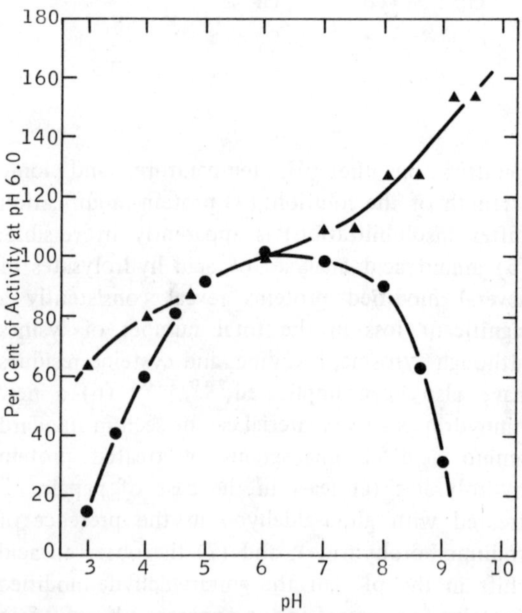


FIGURE 16. pH-Activity curves for a three-layer papain-collodion membrane (▲) and papain (●) using BAEE as substrate. Enzymic activities are expressed as percent of the corresponding esterolytic activities at pH 6.0. (Redrawn from Goldman, R., Kedem, O., Silman, I. H., Caplan, S. R., and Katchalski, E., *Biochemistry*, 7, 486, 1968. With permission.[259])

native and membrane-immobilized papain using BAEE as the substrate. Equally dramatic departures of the pH-activity curves of this crosslinked papain from the native enzyme were observed with *N*-benzoylglycine ethyl ester (BGEE), *a*-*N*-benzoyl-DL-arginine *p*-nitroanilide (BANA), BAA, and acetyl-L-glutamic acid diamide (AGOA) as substrates. The shapes of the pH-activity curves depended on the nature of the products liberated and on the kinetic parameters determining the rate of hydrolysis. These differences were ascribed to the different enzymatically generated microenvironments by the hydrolysis of these substrates. A theoretical treatment of the kinetic behavior[261] of such enzyme-membrane conjugates was also developed and the kinetic behavior of two-enzyme membranes[273] was considered by Goldman and Katchalski. Likewise, Selegny and co-workers[274,275] examined the kinetic behavior of enzyme-membrane systems in considerable detail. The theoretical pH dependence of enzyme activities in membranes of low dielectric constant was also given by Bass and McIlroy.[276] A thorough discussion of these enzyme-membranes can be found in these papers or in several reviews.[1,2,17]

*Michaelis Constant*

Glucose oxidase is the only enzyme whose $K'_m$ has been reported after crosslinking to a water-insoluble derivative. The crosslinked enzyme, impregnated in a cellophane membrane and treated with glutaraldehyde, exhibited a $K'_m$ value identical to the $K_m$ value of the native enzyme (13 m*M*).[263] An insoluble glucose oxidase film prepared by crosslinking the enzyme with glutaraldehyde in the presence of albumin also showed a $K'_m$ value of 13 m*M*.[250] Considerable theoretical discussions of the $K'_m$ of synthetic membrane-bound enzymes have been given.[250,260,261,264,265,273]

*Specificity*

Examples of changes in the specificity of these immobilized enzymes have been reported. Glutamic-aspartic transaminase[249] crosslinked with glutaraldehyde lost its transaminase activity but was still able to form complexes with its antibody. Water-insoluble catalase[277] prepared by crosslinking with glutaraldehyde likewise exhibited changed behavior in its peroxidatic activity. With methanol as the hydrogen donor, the peroxidatic activity decreased elevenfold. On the other hand, the peroxidatic activity decreased eightyfold when the donor was pyrogallol. Schejter and Bar-Eli suggested that this difference in behavior was due to either diminished accessibility of the substrate to the active site of the enzyme or to a hindrance of conformational changes that may be necessary for peroxidatic activity. The immobilized catalase exhibited behavior toward cyanide, hydroxylamine, and aminotriazole similar to that of the native enzyme.

*Stability*

Various stabilities of these water-insoluble enzyme derivatives have been described. Most notably, the thermal stability of such enzymes has been shown to vary from greater to lesser stability relative to the native enzyme. For example, glucose oxidase[263] impregnated in cellophane sheets and crosslinked with glutaraldehyde exhibited superior thermal stability at 37°. On the contrary, papain[259] immobilized as an enzyme-membrane conjugate (the same three-layer papain membrane previously mentioned) showed a decreased thermal stability relative to the native water-insoluble enzyme in the temperature range of 65 to 80°.

The pH stability behavior of the three-layer papain membrane was determined by Goldman et al.[259] The enzymic activity retained by both the three-layer papain membrane and the native enzyme after 10 min at various pH's was similar and quite high in the pH range of 6 to 10. However, the stability of the three-layer papain membrane was greater at lower pH values. Whereas water-soluble unmodified papain retained only 10% of its initial activity on exposure to pH 2.2, the papain collodion membrane retained about 50% of its initial activity after a similar treatment.

The storage stability of several water-insoluble enzyme derivatives has been examined. Papain[237] crosslinked with glutaraldehyde showed no detectable decrease in its esterase activity after 5 weeks at 4°. On the contrary, catalase[277] crosslinked by the same reagent showed a decrease of about 20% in its initial activity after 2 weeks at 4°, after which no further decrease in activity occurred after 5 months of storage. The activity of water-insoluble pepsin[267] prepared by condensation with ethyl chloroformate was reported not to be markedly affected after 6 months of storage.

Still other stabilities reported for these enzyme derivatives are the enhanced mechanical stability of glutaraldehyde-crosslinked carboxypeptidase crystals,[241,242] the enhanced stability toward urea of crosslinked, monolayered trypsin,[253] and the stability of trypsin[166] crosslinked with glutaraldehyde during continuous use for casein digestion.

**Advantages and Disadvantages of the Method**

The single most favorable aspect for using multifunctional reagents as insolubilization agents is that *one* reagent can be utilized to prepare vastly different types of immobilized enzyme derivatives such as crosslinked enzyme gels, enzyme-membranes, adsorbed monolayered derivatives, and water-insoluble polymer-bound conjugates. These bifunctional reagents are available commercially or can be synthesized readily.

The preparation of a water-insoluble enzyme derivative using a multifunctional reagent is simple and gives a derivative that is almost pure protein. These enzyme gel derivatives can be dispersed readily in aqueous solutions. Shortcomings of this particular approach are the need for fairly rigid controls of pH, concentrations, etc. in order to achieve efficient insolubilization, the need for a large amount of protein or coprotein, the often unavoidable inactivation of the enzyme caused by the chemical modification, and the gelatinous nature of these enzyme derivatives which makes it difficult to use them in some column operations.

Water-insoluble enzyme derivatives prepared by adsorption of the enzyme on surface-active, water-insoluble supports followed by intermolecular crosslinking of the enzyme molecules with multifunctional reagents have the advantage that essentially a monolayer of immobilized enzyme molecules can be formed. This offers a large proportion of the immobilized enzyme molecules for possible substrate interaction and transformation to product. Disadvantages of this approach are that experimental conditions must first be determined to ensure good, if not optimum, adsorption of the enzyme on the support, and the necessity to assure that no aggregation of individual colloidal particles occurs. If aggregation occurs and clusters are formed, it will be difficult, if not impossible, to redisperse these into colloidal dispersions.

Low molecular weight multifunctional reagents can also be employed advantageously for extending the reactive group from a preformed, water-insoluble polymer and for preparing water-insoluble enzyme derivatives entrapped within microcapsules.

TABLE 7

**Water-Insoluble Enzyme Derivatives Prepared with Multifunctional Reagents**

| Enzyme immobilized | Multifunctional reagent | Remarks[a] | Ref. |
|---|---|---|---|
| Alcohol dehydrogenase (1.1.1.1) | Glutaraldehyde (?) | Adsorption on cellophane membrane followed by crosslinking (?) | 250 |
| Glucose-6-phosphate dehydrogenase (1.1.1.49) | Glutaraldehyde (?) | Adsorption on cellophane membrane followed by crosslinking (?) | 250 |
| Lactate dehydrogenase (1.1.-.-) | Glutaraldehyde (?) | Adsorption on cellophane membrane followed by crosslinking (?) | 250 |
| Glucose oxidase (1.1.3.4) | Diazobenzidine-3,3'-dianisidine | Adsorption on cellophane membrane followed by crosslinking | 262,264,265 |
| | Glutaraldehyde | Crosslinked with bovine serum albumin | 247 |
| | Glutaraldehyde | Crosslinked with plasma albumin | 250 |

TABLE 7 (Continued)

**Water-Insoluble Enzyme Derivatives Prepared with Multifunctional Reagents**

| Enzyme immobilized | Multifunctional reagent | Remarks[a] | Ref. |
|---|---|---|---|
| Glucose oxidase (1.1.3.4) | Glutaraldehyde | Adsorption on cellophane membrane followed by crosslinking | 250,263 |
| Uricase (1.7.3.3) | Glutaraldehyde (?) | Adsorption on cellophane membrane followed by crosslinking (?) | 250 |
| Catalase (1.11.1.6) | Glutaraldehyde | | 277 |
| | Glutaraldehyde (?) | Adsorption on cellophane membrane followed by crosslinking (?) | 250 |
| | Glutaraldehyde | Adsorption on cheesecloth followed by crosslinking | 100 |
| | Glutaraldehyde | Adsorption on diethylaminoethylcellulose followed by crosslinking | 100 |
| | Glutaraldehyde | Microencapsulation followed by crosslinking | 246 |
| Peroxidase (1.11.1.7) | Glutaraldehyde | Crosslinked with bovine serum albumin | 247,272 |
| Glutamic-aspartic transaminase (2.6.1.1) | Glutaraldehyde | No transaminase activity retained | 249 |
| | Glutaraldehyde | Crosslinked with human albumin or human globulin; no transaminase activity retained | 249 |
| | Ethyl chloroformate | No transaminase activity retained | 249 |
| | Ethyl chloroformate | Condensed with human albumin or human globulin; no transaminase activity retained | 249 |
| | Glutaraldehyde (?) | Adsorption on cellophane membrane followed by crosslinking (?) | 250 |
| Ribonuclease A (2.7.7.16) | 1,5-Difluoro-2,4-dinitro benzene | Whole crystals crosslinked; catalytic activity not reported | 244 |
| | Glutaraldehyde | | 247 |
| Alkaline phosphatase (3.1.3.1) | Glutaraldehyde | Adsorption on collodion membrane followed by crosslinking | 261 |
| α-Amylase (3.2.1.1) | Glutaraldehyde (?) | Adsorption on cellophane membrane followed by crosslinking (?) | 250 |
| Lysozyme (3.2.1.17) | Glutaraldehyde | | 238,247 |
| | Glutaraldehyde | Whole crystals crosslinked; catalytic activity not reported | 245 |
| | Glutaraldehyde | Adsorption on colloidal silica followed by crosslinking | 252 |
| | Ethyl chloroformate | | 266 |
| β-Galactosidase (3.2.1.23) | Glutaraldehyde (?) | Adsorption on cellophane membrane followed by crosslinking (?) | 250 |
| Carboxypeptidase A (3.4.2.1) | Diazobenzidine | Whole crystals crosslinked | 242 |
| | 1,5-Difluoro-2,4-dinitro benzene | Whole crystals crosslinked | 242 |
| | Glutaraldehyde | | 242 |

TABLE 7 (Continued)

**Water-Insoluble Enzyme Derivatives Prepared with Multifunctional Reagents**

| Enzyme immobilized | Multifunctional reagent | Remarks[a] | Ref. |
|---|---|---|---|
| Carboxypeptidase A (3.4.2.1) | Glutaraldehyde | Whole crystals crosslinked | 241-243 |
| | Glutaraldehyde | Adsorption on colloidal silica followed by crosslinking | 159 |
| Pepsin (3.4.4.1) | Ethyl chloroformate | | 267 |
| | Ethyl chloroformate | Condensed with gelatin or amino acids (Lys, Glu) | 267 |
| Trypsin (3.4.4.4) | Glutaraldehyde | | 247,272,276 |
| | Glutaraldehyde | Crosslinked in presence of ammonium sulfate | 166 |
| | Glutaraldehyde (?) | Adsorption on cellophane membrane followed by crosslinking (?) | 250 |
| | Glutaraldehyde | Adsorption on colloidal silica followed by crosslinking | 159,252,253 |
| α-Chymotrypsin (3.4.4.5) | Glutaraldehyde | | 238,272,278 |
| | Glutaraldehyde | Crosslinked with bovine serum albumin | 238 |
| | Glutaraldehyde (?) | Adsorption on cellophane membrane followed by crosslinking (?) | 250 |
| | Glutaraldehyde | Adsorption on colloidal silica followed by crosslinking | 252 |
| | Hexamethylene diisocyanate | | 271 |
| | WRK | | 268 |
| α-Chymotrypsin (*p*-nitrophenyl acyl derivative) | Glutaraldehyde | | 238 |
| α-Chymotrypsin–papain (mercuribenzoate derivative) | Glutaraldehyde | | 238 |
| Chymotrypsinogen | Ethyl chloroformate | | 266 |
| | Glutaraldehyde | | 238 |
| Papain (3.4.4.10) | Diazobenzidine | | 200 |
| | Diazobenzidine | Adsorption on collodion membrane followed by crosslinking | 259 |
| | Diazobenzidine-3,3′-dianisidine | Adsorption on collodion membrane followed by crosslinking | 259 |
| | Diazobenzidine-3,3′-dicarboxylic acid | Adsorption on collodion membrane followed by crosslinking | 259 |
| | Diazobenzidine-2,2′-disulfonic acid | Partially water-soluble | 200 |
| | Diazobenzidine-2,2′-disulfonic acid | Adsorption on collodion membrane followed by crosslinking | 259,260 |
| | Diazobenzidine-3,3′-disulfonic acid | Adsorption on collodion membrane followed by crosslinking | 258 |

TABLE 7 (Continued)

**Water-Insoluble Enzyme Derivatives Prepared with Multifunctional Reagents**

| Enzyme immobilized | Multifunctional reagent | Remarks[a] | Ref. |
|---|---|---|---|
| Papain (3.4.4.10) | 4,4′-Diisothiocyanato-biphenyl-2,2′-disulfonic acid | | 279 |
| | Glutaraldehyde | | 237,238,240, 280,281 |
| Papain (mercuribenzoate derivative) | Glutaraldehyde | | 237,240 |
| Subtilisin Nova (3.4.4.16) | Glutaraldehyde | Crosslinked in presence of ammonium sulfate or acetone | 251 |
| Kallikrein (3.4.4.21) | Glutaraldehyde | Adsorption on colloidal silica followed by crosslinking | 159 |
| Asparaginase (3.5.1.1) | Glutaraldehyde (?) | Adsorption on cellophane membrane followed by crosslinking (?) | 250 |
| | Glutaraldehyde | Microencapsulation followed by crosslinking | 246 |
| Urease (3.5.1.5) | Glutaraldehyde (?) | Adsorption on cellophane membrane followed by crosslinking | 250 |
| | Glutaraldehyde | Adsorption on colloidal silica followed by crosslinking followed by microencapsulation | 255,256 |
| | Glutaraldehyde | Microencapsulation followed by cross-linking | 246 |
| Apyrase (3.6.1.5) | Glutaraldehyde | | 51 |
| | Diazobenzidine (or disulfonic acid derivative) | Adsorption on filter paper followed by crosslinking | 98 |
| | Diazobenzidine (or disulfonic acid derivative) | Adsorption on collodion membrane followed by crosslinking | 98 |
| | Diazobenzidine (or disulfonic acid derivative) | Adsorption on Millipore filter followed by crosslinking | 98 |
| Aldolase (4.1.2.13) | *N,N′*-hexamethylenebis-iodoacetamide | | 282 |
| Carbonic anhydrase (4.2.1.1) | Diazobenzidine-3,3′-dianisidine | Adsorption on cellophane membrane followed by crosslinking | 262 |
| | Glutaraldehyde (?) | Adsorption on cellophane membrane followed by crosslinking (?) | 250 |
| | Glutaraldehyde | Adsorption on Silastic (Dow-Corning 500-3 Silastic sheet) concomitant with cross-linking | 257 |

TABLE 7 (Continued)

**Water-Insoluble Enzyme Derivatives Prepared with Multifunctional Reagents**

| Enzyme immobilized | Multifunctional reagent | Remarks[a] | Ref. |
|---|---|---|---|
| Phenylalanine decarb-oxylase (4.-.-.-) | Glutaraldehyde (?) | Adsorption on cellophane membrane followed by crosslinking (?) | 250 |
| Tyrosine decarboxylase (4.-.-.-) | Glutaraldehyde (?) | Adsorption on cellophane membrane followed by crosslinking (?) | 250 |
| Triose-phosphate isomerase (5.3.1.1) | Glutaraldehyde (?) | Adsorption on cellophane membrane followed by crosslinking (?) | 250 |
| Glucose-6-phosphate isomerase (5.3.1.9) | Glutaraldehyde (?) | Adsorption on cellophane membrane followed by crosslinking (?) | 250 |

[a]Unless otherwise noted, the water-insoluble enzyme derivative was prepared by the reaction of an aqueous solution of the enzyme with the multifunctional reagent.

# Chapter 5

# ADSORPTION

The adsorption of proteins onto surfaces has been extensively investigated, particularly with enzymes. Adsorption of enzymes can occur at air-water, water-oil, and water-solid interfaces. The main emphasis of most studies on the adsorption of enzymes has been on the nature of the adsorption process or on purification but not on changed chemical and physical properties or on the possibility of using these immobilized enzymes in continuous catalytic processes. Furthermore, even when the activity of an adsorbed enzyme was reported, the main interest of the investigation was not continuous catalysis. Especially in earlier studies in which the activity of an "adsorbed" enzyme is reported, it is extremely difficult to ascertain whether the activity was indeed due to adsorbed enzyme or to some desorbed, soluble enzyme. Little information is given in these studies about the integrity of the enzyme-adsorbent conjugate. Only within the last two decades has the adsorption of enzymes onto water-insoluble materials been used commonly as a method for the immobilization of these molecules for use in continuous catalytic processes.

This chapter deals only with water-solid interfacially adsorbed enzymes. The discussion is also restricted to systems in which the main interest of the investigator(s) was the immobilization of the enzyme for a continuous catalytic process or where the activity of an enzyme-adsorbent conjugate was reported. The adsorption of enzymes at other interfaces[19,283] and their adsorption for purification purposes[284] have been reviewed. The reviews by James and Augenstein[283] and McLaren and Packer[19] present excellent discussions of adsorbed enzymes.

This chapter also discusses the immobilization of enzymes by complex formation with water-soluble, charged molecules. Immobilization of enzymes by adsorption onto water-insoluble matrices followed by intermolecular crosslinking is discussed in Chapter 4.

**Adsorption of Enzymes**

This method for preparing water-insoluble enzyme conjugates is certainly one of the simplest. It consists of contacting an aqueous solution of an enzyme with a surface-active adsorbent and washing the resulting conjugate to remove any nonadsorbed enzyme. Many different enzymes and adsorbents have been employed (see Table 9). The more commonly employed water-insoluble adsorbents are listed in Table 8. The adsorbents are either organic or inorganic in nature and often require special pretreatments in order to insure good adsorption. They are, however, commercially available.

The adsorption of an enzyme onto a water-insoluble material is dependent on such experimental variables as pH, nature of the solvent, ionic strength, concentration of protein and adsorbent, and temperature. Good discussions of these variables have been given by Zittle,[284] James and Augenstein,[283] McLaren and co-workers,[19,285,286] and Hummel and Anderson.[287] It is extremely important to be cognizant of these factors and to control them for optimum adsorption and retainment of activity. Activity and adsorption do not necessarily go hand in hand.

The mechanism of the adsorption process is often multivariant and it is at times difficult to differentiate clearly between the various possibilities. Adsorption of enzymes onto water-insoluble matrices can be attributed to an ion-

TABLE 8

**Commonly Employed Adsorbents for the Immobilization of Enzymes**

| |
|---|
| Alumina |
| Anion-exchange resins |
| Calcium carbonate |
| Carbon |
| Cation-exchange resins |
| Celluloses |
| Clays |
| Collagen |
| Collodion |
| Conditioned metal or glass plates |
| Diatomaceous earth |
| Hydroxylapatite |

exchange mechanism, to simple physical adsorption at the external surface of a particle, or to "physicochemical bonds" created by hydrophobic interactions, van der Waals attractive forces, etc.[285,288–293] The mode of adsorption will depend greatly on the nature of the water-insoluble carrier.

**Properties**

The immobilization of an enzyme by adsorption is a physical method for localizing an enzyme. In principle, no permanent change in either a chemical or physical property is expected after the enzyme has once again been eluted from the adsorbent. On the adsorbent, however, considerable changes can be expected due either to characteristic changes in its intrinsic properties (such as conformational changes or charge elimination) or to microenvironmental effects caused by the charge of a carrier, diffusion, or steric repulsions.

*Activity*

The activity of adsorbed enzymes can vary anywhere from nil to fairly high values. In some favorable cases, the relative activity of these immobilized enzymes even approaches that of the soluble enzymes.[294] Often, however, the activity of the conjugates is only qualitatively reported as either being active or not. Relative activities have been reported for (the relative activity in percent is given in parentheses): glucose-6-phosphate dehydrogenase[294] on collodion (50), invertase[130] on bentonite (1.2), leucine aminopeptidase[157] on calcium phosphate gel (78), polynucleotide phosphorylase[295] on nitrocellulose (30), deoxyribonuclease[296] on cellulose powder (65 to 90), and α-chymotrypsin[297] on kaolinite (70). DEAE-cellulose immobilized enzymes exhibited similar varied activities: β-fructofuranosidase (25 to 31)[298] (30 to 50)[299] (50),[300] ATP deaminase[301] (ca. 25), glucoamylase[302] (16 to 58), and catalase[303] (ca. 70). Aminoacylase[291] adsorbed on DEAE-Sephadex exhibited 86% activity when compared with its soluble counterpart. Binding capacities of some enzymes on several supports have been reported also in studies dealing with immobilized enzymes.[157,289,298,304–306]

*pH-Activity Behavior*

The pH-activity behavior of water-insoluble enzyme-adsorbent conjugates is analogous to the behavior observed with covalently bonded water-insoluble derivatives or with other carrier-bound enzymes. The pH-activity curves of the freely soluble enzymes can be displaced either toward more alkaline or acid pH's. Likewise, the shape and optimum pH of the water-insoluble conjugates may not be changed at all. Displacements in either direction have been attributed to the microenvironment created by the charged carriers. In fact, McLaren and co-workers[307–309] predicted and observed this pH-activity curve displacement phenomenon for these water-insoluble enzyme conjugates well before the same behavior was observed for covalently bonded enzyme derivatives. McLaren[309] also presented a mathematical treatment of the pH difference between the environment surrounding a polyelectrolyte carrier and the bulk phase of the solution.

Shifts toward more alkaline pH's were observed for α-chymotrypsin[307–309] adsorbed on kaolinite and for glucoamylase[310] adsorbed on acid clay. The latter system also exhibited a somewhat narrower pH-activity profile than the native enzyme. Displacements of the pH-activity curve toward more acid pH's were observed for glucoamylase[310] on DEAE-Sephadex A-25, ATP deaminase[301] on DEAE-cellulose, aminoacylase[311] on DEAE-cellulose, and invertase[299,300] on DEAE-cellulose. In the case of aminoacylase on either support, the acid pH-activity displacement was noted to be substrate-dependent. For example, the optimum pH for the hydrolysis of *N*-acetyl-DL-methionine was 7.0 and 7.5 for the Sephadex-conjugate and free enzyme, respectively. Other pH optima for the immobilized and native aminoacylase were 6.0 and 7.5 for *N*-acetyl-DL-phenylalanine, 6.0 and 7.5 for *N*-acetyl-DL-tryptophan, and 6.0 and 7.0 for *N*-acetyl-DL-valine. On the other hand, identical pH optima for the adsorbed and native enzymes were observed for *N*-chloroacetyl derivatives of phenylalanine, tryptophan, and tyrosine. Besides these examples of unaltered pH optima for the native and immobilized enzymes, the same optimum was noted for urease[312,313] adsorbed on kaolinite and for soluble urease.

*Michaelis Constant*

The value of the apparent Michaelis constant for adsorbed enzymes has been observed to be higher, lower, or the same as that of the corresponding soluble enzymes. Higher $K_m$ values were observed for malate dehydrogenase[314] adsorbed on silica gel, glucoamylase[310,315] on activated charcoal, aminoacylase[316] on DEAE-Sephadex, and aminoacylase[311] on DEAE-cellulose. Lower values were observed for aminoacylase[311,316] on DEAE-cellulose or DEAE-Sephadex and for malate dehydrogenase[314] on silica gel. The apparent difference in the behavior of these same immobilized enzyme systems can be attributed to the use of different substrates. Identical constants (relative to the soluble enzymes) were observed for acid phosphatase[317] and phosphoglucomutase[317] on silica gel. An expanded discussion of the Michaelis constant of immobilized enzymes is given in Chapter 3.

*Specificity*

Some interesting substrate and inhibitor specificities have been reported. For example, dextransucrase[318] adsorbed on DEAE-Sephadex exhibited a changed substrate specificity relative to the soluble enzyme. The polymer formed by the insoluble enzyme had a higher glycogen value and fructose content. ATP deaminase[301] showed the same order of reactivity toward ATP, ADP, and AMP as did the native enzyme, but the relative ratio of activities was different. A surprising activation of aminoacylase adsorbed on DEAE-Sephadex or DEAE-cellulose was observed by Tosa et al.[316,319] Incubation of the enzymes for 1 hr in 6 *M* urea and then assaying for activity in 2 *M* urea showed enhanced catalysis for these conjugates relative to their initial activities. The activation of the immobilized enzymes by urea was attributed to the conversion of the rigid structures of the conjugates by the denaturant to more flexible and active conformers.

Additional examples of changed behavior of immobilized enzymes are the enhanced resistance of β-fructofuranosidase[298] toward mercuric acetate inhibition, the metal inactivation of aminoacylase,[311] and the greater inhibition of adsorbed glucoamylase[320] by mercuric acetate.

*Stability*

A considerable amount of information is available on the various types of stabilities of adsorbed water-insoluble enzyme conjugates. As in the case with covalently bonded water-insoluble enzyme derivatives, a spectrum of stabilities exists.

The thermal stability of adsorbed enzymes can be increased, decreased, or not changed at all. Enhanced thermal stability was found for phosphomonoesterase[321] adsorbed on CM-cellulose, aminoacylase[316] on DEAE-Sephadex, NAD pyrophosphorylase[322] on hydroxylapatite, and aminoacylase[311] on DEAE-cellulose. On the contrary, decreased thermal stability was observed for glucose-6-phosphate dehydrogenase[294] adsorbed on collodion, β-fructofuranosidase[298,300] on DEAE-cellulose, leucine aminopeptidase[157] on calcium phosphate gel, glucoamylase[320] on activated clay, and ATP deaminase[301] on DEAE-cellulose. Essentially similar thermal stability (relative to soluble enzyme) was observed for urease[312] adsorbed on kaolinite.

Storage stabilities for adsorbed enzymes have also been reported.[157,306,316,318] β-Fructofuranosidase[298] adsorbed on DEAE-cellulose lost negligible activity after 15 weeks of storage at 4°, the stability of urease[312] on kaolinite was similar to that of the native enzyme at 2°, and polynucleotide phosphorylase[295] on nitrocellulose had a halflife of one week at 4°.

Other stabilities that have been reported are the enhanced proteolytic stability (toward pronase and trypsin hydrolysis) of aminoacylase[316] adsorbed on DEAE-Sephadex, the diminished pH stability (at pH 5.2) of invertase[299] on DEAE-cellulose and glucoamylase[320] on activated clay, and the enhanced pH stability of urease[313] on kaolinite. The stability of adsorbed enzymes has also been investigated in continuous catalytic processes[302,322-324] and upon x-ray radiation.[296]

**Reversibility of Adsorption**

An important theoretical as well as practical consideration of these immobilized enzymes is the reversibility of adsorption. In principle, the physical adsorption of a protein onto a surface should be a reversible process. A change in pH, ionic strength, temperature, etc. of the mixture should cause the bound enzyme to desorb from the water-insoluble matrix and be released into the aqueous solution phase. In practice, however, this may not be feasible. Desorption of an enzyme may not be possible, either because some irreversible process involving the carrier and enzyme has occurred or the appropriate conditions for desorption have not yet been found. Irreversible binding has been reported for urease[312] adsorbed on kaolinite and for lactate dehydrogenase[295] on nitrocellulose.

Irreversible binding is not a problem if the immobilized enzyme is still active and if it is to be used in a continuous process. In fact, it may be quite desirable to have strong binding of an enzyme, for weak binding can be a serious problem leading to desorption ("leakage") of the enzyme with subsequent loss of activity and contamination of the product(s). Such a problem of desorption was observed for glucoamylase[310, 315] on activated charcoal or acid clay. Likewise, the desorption behavior of other enzyme-adsorbent conjugates has been reported and the following examples will illustrate the complexity of this process. A delicate balance of experimental conditions must be maintained in order to either retain or elute the enzyme. Some examples of substrate-induced desorption (dependent on concentration) are glucose-6-phosphate dehydrogenase[294] adsorbed on collodion, aminoacylase[292,323] on DEAE-cellulose, aminoacylase[325] on DEAE-Sephadex, and amylase[326] on acid clay. pH-Induced desorption was found for papain[306] adsorbed on glass, leucine aminopeptidase[157] on calcium phosphate gel, and catalase[303] on DEAE-cellulose. Catalase adsorbed on DEAE-cellulose exhibited somewhat unusual desorption behavior in that it could be eluted either with a change in pH or ionic strength but not with its substrate. Other examples of ionic strength-induced desorption of enzymes are leucine aminopeptidase[157] adsorbed on calcium phosphate gel and ATP deaminase[301] on DEAE-cellulose. Desorption of enzymes can thus be achieved by changing solution variables such as pH, ionic strength, concentration of substrate, etc., individually (holding all others constant) or simultaneously.

If adsorption of an enzyme is reversible, then, in principle, it should be feasible to regenerate the support and reuse it for immobilizing either more of the same enzyme or different enzymes. Tosa et al.[324] demonstrated reactivation and reuse of an aminoacylase-DEAE-Sephadex column. Reactivation was accomplished either by treatment with sodium hydroxide followed by addition of more aminoacylase or simply by adding additional enzyme.

Many of the adsorbents discussed in the text and listed in Table 8 are used for the purification of enzymes. Thus, it is interesting to speculate whether purification and immobilization could not be achieved simultaneously in cases of crude enzyme mixtures. This indeed seems to be possible. For example, Nikolaev and Mardashev[327] obtained a purification of asparaginase from a crude homogenate adsorbed on CM-cellulose. Sundaram and Crook[312] likewise suggested the possibility of this occurring in their system of urease adsorbed on kaolinite. Even in homogeneous enzyme preparations, preferential adsorption and desorption could be used for different conformers of the same enzyme (and still active) or for inactive and active forms.[283,328]

**Advantages and Disadvantages**

Advantages of this method for immobilizing enzymes are simplicity, the large choice of differently charged and shaped carriers, and the possibility of achieving simultaneous purification and immobilization. The adsorption of enzymes onto water-insoluble supports is an exceedingly simple method operationally. It is "mild" and often causes little or no overall enzymic inactivation. In principle, the method is completely reversible, permitting reuse of both the support and the catalyst for other purposes. A number of differently charged and surface-treated supports are available; different physical forms of the carriers (sheets, fibers, etc.) also exist.

Disadvantages of the method exist. Although it is operationally easy to prepare such water-insoluble enzyme conjugates, optimum conditions for achieving this are often a matter of trial and error. The correct conditions of pH, ionic strength, and temperature must be determined in order to achieve and maintain good adsorption and activity. Also, if strong binding between the enzyme and the support does not exist, desorption of the protein will occur with subsequent loss of catalytic activity and contamination of products. "Leakage" problems of immobilized enzymes can be a nuisance, especially when operating at high substrate concentrations.

**Immobilization by Complex Formation With Water-Soluble Agents**

Recently, Negoro[329] reported a method for preparing water-insoluble saccharase by complexing the soluble enzyme with a water-soluble complexing agent to give an enzymatically active, water-insoluble enzyme conjugate. The water-insoluble complex was prepared by adding a solution of tannic acid – a polymeric glycoside derivative of gallic acid – to a solution of the enzyme. The precipitate that formed was isolated by centrifugation and washed with water.

No significant differences were observed between the native saccharase and the insoluble complex with regard to the optimum pH and temperature, but the pH-activity and activity-temperature curves of the complex exhibited a different shape. The complex was more active at higher pH's and at higher temperatures but less active in the acid region below 4. Dissociation of the enzyme-tannic acid complex occurred at low pH's. Enhanced resistance toward pronase hydrolysis was also observed for the water-insoluble complex. The complex was used in a packed-bed reactor for the continuous hydrolysis of sucrose.

Although examples of "precipitation" or complex formation between water-soluble enzymes and various reagents have been known[330,331] for some time, such water-insoluble materials had not been employed previously for continuous catalytic conversions.

TABLE 9

**Water-Insoluble Enzyme-Adsorbent Conjugates**

| Enzyme immobilized | Adsorbent | Ref. |
|---|---|---|
| Lactate dehydrogenase (1.1.1.27) | Millipore filter (nitrocellulose) | 295 |
| Malate dehydrogenase (1.1.1.37) | Silica gel | 314 |
| | Silica gel coated with monolayer of lecithin, cephalin or cholesterol | 314 |
| Glucose-6-phosphate dehydrogenase (1.1.1.49) | Collodion | 294 |
| | Silica gel | 332 |
| | Silica gel coated with monolayer of lecithin | 332 |
| Glucose oxidase (1.1.3.4) | Glass | 305 |
| Glucose oxidase – catalase | Alumina | 333 |
| | Amberlite CG-50 Type II | 333 |
| | Carbon | 333 |
| | Clay | 333 |
| Succinate dehydrogenase (1.3.99.1) | Silica gel coated with monolayer of cephalin or lecithin | 323 |
| | Carbon coated with monolayer of cephalin or lecithin | 323 |
| Catalase (1.11.1.6) | Bentonite | 334 |
| | Calcium carbonate | 334 |
| | Carbon coated with lauric acid or cephalin | 335 |
| | Carboxylic acid ion-exchange resin Amberlite XE-97 | 38 |
| | Cellulose, carboxymethyl ether | 303 |
| | Kaolinite | 334 |
| | Metal or glass plates treated with barium stearate, thorium nitrate, and sodium deoxycholate | 331,336 |
| | Polyaminostyrene | 38 |
| | Silica coated with lauric acid, cephalin, or tri-dodecylamine | 335 |

TABLE 9 (Continued)

**Water-Insoluble Enzyme-Adsorbent Conjugates**

| Enzyme immobilized | Adsorbent | Ref. |
|---|---|---|
| Dextransucrase (2.4.1.5) | Sephadex, diethylaminoethyl ether | 318 |
| Hexokinase (2.7.1.1) | Silica gel | 337,338 |
| | Silica coated with lauric acid or cephalin | 338 |
| Phosphoglucomutase (2.7.5.1) | Carbon coated with cephalin | 317 |
| | Silica | 317 |
| | Silica coated with cephalin | 317 |
| NAD pyrophosphorylase (2.7.7.1) | Hydroxylapatite | 322 |
| Polynucleotide phosphorylase (2.7.7.8) | Millipore filter (nitrocellulose) | 295 |
| Ribonuclease A (2.7.7.16) | Cationic resin SBS 4 (H) | 339 |
| | Dowex®-2 anion-exchange resin | 340 |
| | Dowex-50 cation-exchange resin | 340 |
| | Glass | 289,340 |
| Lipase (3.1.1.3) | Carboxylic acid ion-exchange resin Amberlite XE-97 | 38 |
| | Polyaminostyrene | 38 |
| Acid phosphatase (3.1.3.2) | Carbon coated with cephalin | 317 |
| | Silica | 317 |
| | Silica coated with cephalin | 317 |
| Phosphomonoesterase (3.1.3.-) | Cellulose, carboxymethyl ether | 321 |
| Deoxyribonuclease I (3.1.4.5) | Cellulose | 296,341 |
| α-Amylase (3.2.1.1) | Bentonite | 334 |
| | Collagen | 342 |
| | Kaolinite | 334 |
| Amylase (α?) | Alumina | 326 |
| | Amberlite CG-50 | 326 |
| | Calcium phosphate gel | 326 |
| | Carbon | 326 |
| | Clay (acid) | 326 |
| | Diatomaceous earth | 326 |
| | Dowex 1-X4 | 326 |
| | Silica gel | 326 |
| β-Amylase (3.2.1.2) | Bentonite | 334 |
| | Collagen | 342 |
| | Kaolinite | 334 |
| Glucoamylase (3.2.1.3) | Amberlite CG-4B Type II | 333 |
| | Amberlite IR45 ($OH^-$) | 302 |
| | Carbon | 310,315 |
| | Cellulose, carboxymethyl ether | 333 |
| | Cellulose, diethylaminoethyl ether | 302,333,343 |
| | Clay (acid) | 310,315,320 |
| | Diatomaceous earth | 310 |

TABLE 9 (Continued)

**Water-Insoluble Enzyme-Adsorbent Conjugates**

| Enzyme immobilized | Adsorbent | Ref. |
|---|---|---|
| Glucoamylase (3.2.1.3) | Dowex-1-X10 ($Cl^-$) | 302 |
| | Sephadex, carboxymethyl ether | 333 |
| Cellulase (3.2.1.4) | Cellulose | 304 |
| | Collagen | 342 |
| Lysozyme (3.2.1.17) | Collagen | 342,344 |
| β-Galactosidase (3.2.1.23) | Cellulose, diethylaminoethyl ether | 64 |
| β-Fructofuranosidase (3.2.1.26) | Aluminum hydroxide (gelatinous) | 345-347 |
| | Bentonite | 130 |
| | Carbon | 345-347 |
| | Cellulose, diethylaminoethyl ether | 298-300 |
| | Collagen | 342,344 |
| Leucine aminopeptidase (3.4.1.1) | Calcium phosphate gel | 157 |
| Pepsin (3.4.4.1) | Cellulose, diethylaminoethyl ether | 290 |
| | Metal or glass plates treated with barium stearate, thorium nitrate, and sodium silicate or sodium deoxycholate | 336,348 |
| Trypsin (3.4.4.4) | Cellulose, carboxymethyl ether | 290 |
| | Cellulose nitrate | 290 |
| | Cellulose phosphate | 290 |
| | Kaolinite | 297 |
| | Glass[a] | 289 |
| Trypsin (acetyl derivative) | Kaolinite | 297 |
| α-Chymotrypsin (3.4.4.5) | Cellulose, carboxymethyl ether | 290 |
| | Cellulose nitrate | 290 |
| | Cellulose phosphate | 290 |
| | Glass[a] | 289 |
| | Kaolinite | 297,307-309 |
| | Metal-coated glass plates coated with barium stearate (and also treated with uranyl acetate) | 349,350 |
| Papain (3.4.4.10) | Glass | 305,306 |
| Proteases (3.4.-.-) | Cellulose, diethylaminoethyl ether | 290 |
| Asparaginase (3.5.1.1) | Cellulose, carboxymethyl ether | 327,351 |
| | Cellulose, diethylaminoethyl ether | 351 |
| Urease (3.5.1.5) | Collagen | 293,342 |
| | Kaolinite | 312,313 |
| | Metal or glass plates treated with barium stearate, thorium nitrate, and sodium silicate or sodium deoxycholate | 336,348 |

TABLE 9 (Continued)

**Water-Insoluble Enzyme-Adsorbent Conjugates**

| Enzyme immobilized | Adsorbent | Ref. |
|---|---|---|
| Aminoacylase (3.5.1.14) | Alumina (acidic and neutral) | 291 |
| | Amberlite IRC-50[a] | 291 |
| | Amberlite IR-4B and IR-45 ($Cl^-$)[a] | 291 |
| | Carbon[a] | 291 |
| | Cellulose, diethylaminoethyl ether | 290-292,311,323,324 |
| | Cellulose, epichlorohydrin triethanolamine ether | 291 |
| | Cellulose, triethylaminoethyl ether | 291 |
| | Diaion® SA-11A and SA-21A ($Cl^-$) | 291 |
| | Sephadex, carboxymethyl ether C-50[a] | 291 |
| | Sephadex, diethylaminoethyl ether | 291,316,319,324, 325,352 |
| | Sephadex, sulfoethyl ether C-50[a] | 291 |
| ATP deaminase (3.5.4.-) | Carbon | 301 |
| | Cellulose, diethylaminoethyl ether | 301 |
| | Cellulose, triethylaminoethyl ether | 301 |
| ATPase (3.6.1.3) | Millipore or Sartorius filters | 98 |
| Lysine decarboxylase (4.1.1.18) | Alumina | 353 |
| D-Oxynitrilase (4.1.2.10) | Cellulose, epichlorohydrin triethanolamine ether | 354 |
| Glucose isomerase (5.3.1.-) | Collagen | 342 |

[a]Adsorption but no activity.

Chapter 6

# ENTRAPMENT WITHIN CROSSLINKED POLYMERS

Enzymes can be immobilized by entrapment within the interstitial space of crosslinked water-insoluble polymers. The method involves the formation of a highly crosslinked network of a polymer in the presence of an enzyme. Enzyme molecules are physically entrapped within the polymer lattice and cannot permeate out of the gel matrix, but appropriately sized substrate and product molecules can transfer across and within this network to insure a continuous transformation (see Figure 17). Other commonly used names for the method are inclusion, lattice entrapment, and occlusion.

Lattice entrapment was first employed successfully for the immobilization of trypsin, α-chymotrypsin, and other enzymes by Bernfeld and Wan[355] in 1963. Since then, numerous enzymes have been immobilized in this manner, and several polymeric networks have been used (see Table 10). The most commonly employed crosslinked polymer for enzyme entrapment is the well-known polyacrylamide gel system, but silicone rubber (Silastic®), starch, and silica gel have also been used. Dickey[356] reported a partially successful entrapment of urease and catalase in silica gel as early as 1955. Although immobilization of enzymes in these systems is visualized to occur by entrapment within the lattice structure of the polymer, part of the immobilization could be due to physical adsorption. This is especially so with charged polymers such as silica gel.

## Nature of Crosslinked Polymeric Matrices

The polyacrylamide gel system is produced by the reaction of acrylamide and *N,N'*-methylenebisacrylamide. The polymerization reaction can be initiated in several ways and that used most often for enzyme immobilization is shown in Equation 52. The procedure for the formation of the crosslinked polyacrylamide-enzyme conjugate is identical to that used for the preparation of polyacrylamide gels employed commonly for separation and isolation of enzymes, except that in this case the protein is present during the polymerization. A recent review of polyacrylamide gel electrophoresis by Chrambach and Rodbard[357] provides a quick survey of the more important parameters of the reaction and the structure of the produced gel. Several studies dealing with polyacrylamide gel-immobilized enzymes have also described some pertinent details about this

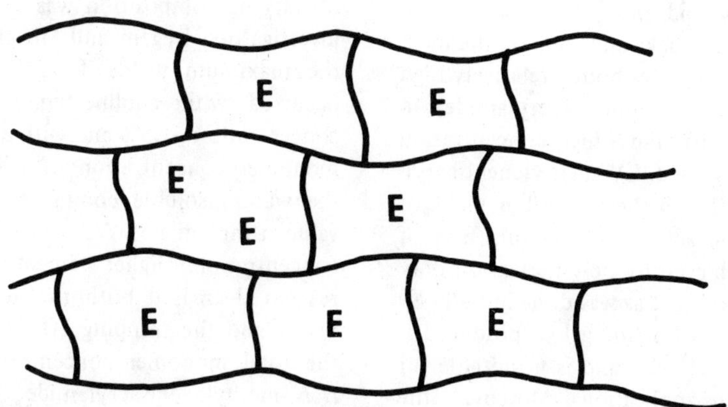


FIGURE 17. Cross-sectional view of lattice-entrapped enzyme conjugate showing polymer chains and occluded enzyme molecules.

$$
\underset{\substack{|\\ \text{CONH}_2}}{\text{CH}_2=\text{CH}} + \underset{\substack{|\\ \text{CO}\\ |\\ \text{NH}\\ |\\ \text{CH}_2\\ |\\ \text{NH}\\ |\\ \text{CO}\\ |\\ \text{CH}_2=\text{CH}}}{\text{CH}_2=\text{CH}} \xrightarrow[\text{TEMED}]{\text{K}_2\text{S}_2\text{O}_8} -\text{CH}_2-\underset{\substack{|\\ \text{CONH}_2}}{\text{CH}}-\text{CH}_2-\underset{\substack{|\\ \text{CO}\\ |\\ \text{NH}\\ |\\ \text{CH}_2\\ |\\ \text{NH}\\ |\\ \text{CO}\\ |\\ -\text{CH}_2-\text{CH}-}}{\text{CH}}- \qquad (52)
$$

procedure. The total concentration and relative ratio of acrylamide and *N,N'*-methylenebisacrylamide determine the pore size of the interstitial space within which the enzyme molecules are entrapped and the physical nature of the water-insoluble material produced.

Hicks and Updike[358] examined the characteristics and activity of lactate dehydrogenase immobilized in different compositions of acrylamide and *N,N'*-methylenebisacrylamide and observed that the best mechanical rigidity was obtained at higher gel concentrations (total acrylamide and *N,N'*-methylenebisacrylamide). However, at any one total concentration, an increase in the relative amounts of crosslinking reagent decreased the mechanical rigidity of the gel but gave a higher yield of immobilized enzyme activity per unit of soluble enzyme activity introduced before polymerization. For example, at a total concentration of 5% (acrylamide and *N,N'*-methylenebisacrylamide), a 5% *N,N'*-methylenebisacrylamide concentration (95% acrylamide) produced a gel of "excellent" mechanical rigidity which showed a relative activity of 60%. The polyacrylamide gel composed of 10% *N,N'*-methylenebisacrylamide exhibited only "fair" rigidity and showed a 66% relative activity. According to Hicks and Updike, the most suitable gel material requires both a relatively high concentration of the monomer (acrylamide) to give mechanical rigidity and a high concentration of crosslinking reagent (*N,N'*-methylenebisacrylamide) to achieve the highest possible yield of immobilized enzyme activity. Immobilization of enzymes can be achieved by polymerization only of the crosslinking reagent as was done initially by Bernfeld and Wan,[355] but the gel so produced is very soft, sediments slowly, and is unsuitable for use in flow system applications. Moreover, the degree of concentration of the crosslinking reagent is severely limited by its solubility in water, which is approximately 3%. The appearance of polyacrylamide gels varies from clear to opaque, depending on the exact composition.

Hicks and Updike[358] also examined the nature of the polymerization catalyst. They noted that mixtures with a high percentage of monomer polymerize more effectively with persulfate, and that solutions with a higher percentage of crosslinking reagent polymerize better with riboflavin and a photocatalyst. Often, it is best to have both catalytic systems present. Other catalytic systems employed for polymerization during the immobilization of enzymes have been TEMED and persulfate,[107,355,359,360] β-dimethylaminopropionitrile and persulfate,[106,361-364] and x-ray radiation.[365] The latter initiation method has, in principle, the inherent advantage of being nonchemical in nature, allowing better heat control and permitting quick termination and control of the initiation step. The polymerization of acrylamide and *N,N'*-methylenebisacrylamide is usually conducted at room temperature or somewhat lowered temperatures and in the absence of atmospheric oxygen. The oxygen molecule, being a paramagnetic species, is a very potent polymerization inhibitor.

The trapping efficiency of polyacrylamide gels of varying composition was examined by several investigators. Degani and Miron[364] observed that the maximum yield of activity trapping (56%) occurred with cholinesterase at a crosslinking concentration of 5% and with a total and constant monomer concentration of 15%. The activity of the water-insoluble conjugate reached its highest value at the same 5% *N,N'*-methylenebisacrylamide concentration. Higher percentages of crosslinking reagent decreased both the activity of the conjugate and the trapping efficiency. The effect of the total monomer concentration (at a constant *N,N'*-methylenebisacrylamide concentration of 5%) on the activity of the conjugate and the trapping efficiency was likewise investigated. With increasing concentrations of total monomer, the yield of protein trapping increased as expected by the decrease in gel porosity. However, the activity of the conjugate and the yield of activity trapping reached a maximum at 15% total monomer concentration. Higher total monomer concentrations decreased both the activity and the yield of activity trapping. Degani and Miron attributed this decrease in activity to inactivation of the enzyme occurring in the polymerization mixture. Acrylamide apparently acts as a denaturing agent on cholinesterase in a manner similar to that observed with urea.

Strandberg and Smiley[366] examined the entrapment of glucose isomerase with varying *N,N'*-methylenebisacrylamide concentrations (constant total monomer) and observed a similar trend of lower trapping with increasing concentrations of crosslinking reagent. The typical trapping efficiency of the polyacrylamide for glucose isomerase was roughly 40 to 50% of the enzyme added to the system. About 30% of the entrapped enzyme exhibited activity in the normal assay.

Bernfeld et al.[360,367] examined the entrapment and catalytic efficiency of a $^{14}C$-labeled aldolase-*N,N'*-methylenebisacrylamide water-insoluble conjugate prepared without added acrylamide. The water-insoluble conjugate prepared in this manner contained 55% of the radioactivity and exhibited 10.4% of the total enzymic activity. The aqueous solution contained 44.2% of the radioactivity and exhibited 33.1% of the enzymic activity. The ratio of enzyme activity to radioactivity in the aqueous phase which remained after termination of the polymerization was reduced by only 25%. On the contrary, the ratio of activity to radioactivity for the insoluble enzyme-conjugate was reduced by about 80%. These results indicated that four times more enzyme protein than enzymic activity was associated with the water-insoluble carrier and Bernfeld et al.[360] reasoned that the most plausible explanation of the findings was that only a portion of the entrapped aldolase is enzymatically active. Only the entrapped aldolase near the polymer-water interface apparently exhibited activity.

As mentioned, Silastic resin (silicone rubber),[362,368-370] silica gel,[356,371] and starch gel[372-374] have been used for lattice entrapment of enzymes. Silastic resin (Dow Corning Co., Midland, Mich.) has the general chemical structure shown below where n is in the order of 10,000. The resin also contains a silica filler, and on addition of the catalyst, stannous octoate, water-insoluble silicone rubber is produced. The water-insoluble enzyme conjugate is prepared by adding the enzyme to an excess of the Silastic resin

$$\left( -\underset{CH_3}{\overset{CH_3}{\underset{|}{\overset{|}{Si}}}}-O- \right)_n \qquad \text{Silastic}^{\circledR} \text{ resin}$$

(100-fold w/w excess), stirring the mixture for several minutes, and then adding the stannous octoate in catalytic portions. Gel formation takes place within half an hour under normal circumstances, and the product is a rigid material that can be cut into any desired shape. Reasonable good entrapment efficiency seems to have been observed with trypsin and chymotrypsin.[362] However, Guilbault and Das[370] have reported some unfavorable characteristics of Silastic-immobilized enzyme derivatives. They noted that the polymerization is "rigorous" and that up to 80% of the enzymic activity is lost in the immobilization. In addition, the silicone rubber pads cracked upon drying and did not become uniform upon rehydration because of the hydrophobic nature of the Silastic resin.

Starch gels have been employed by Guilbault and colleagues[372-374] for the preparation of urethane foam-supported, immobilized enzyme pads. These pads[372] are prepared by pouring a slurry of Connaught-type starch into a boiling mixture of buffer and glycerine, heating the resulting mixture until a clear solution is obtained, and then cooling it to 47°. At this point, a solution of the enzyme is added to the clear starch solution with stirring for several seconds; the resulting mixture is immediately poured onto open-cell urethane foam. The enzyme-starch solution is gently worked into the urethane foam, excess liquid is removed by squeezing, and the pad is cooled in a refrigerator for an hour in order to form the permanent gel. The large pad can be cut into smaller segments. This method of immobilization was reported only for cholinesterase. The addition of glycerine to the starch gel produces pads that are less subject to mechanical damage and which are able to rehydrate more quickly than pads without glycerine. Other additives (Triton® X-100 and Kraystay® K) are more superior storage additives than glycerine.[374]

Several enzymes, most notably trypsin, have also been immobilized by incorporation within a polymerizing silicic acid sol.[356,371] The use of this procedure for enhancing the storage stability of certain materials (including enzymes) was suggested by Dickey[356] in 1955. Urease and catalase immobilized in this manner exhibited some detectable activity, but muscle adenylic acid deaminase was completely inactivated by the condition used to form the gel. Only recently has this method been used successfully with an enzyme. Johnson and Whateley[371] immobilized trypsin within the lattice structure of a silica "xerogel." The term "xerogel" was used by them to mean a dried-out nonswelling structure such as commercial silica gel. The term "hydrogel" was used for a water-rich colloidal system with a finite, rather small yield stress. The procedure for preparing the entrapped trypsin was as follows: the pH of the silicic acid sol was adjusted with sodium hydroxide to be between 6 and 7, a solution of the enzyme was added to the sol, and after adjusting the NaCl concentration, the sol was allowed to set to a hydrogel. Gel formation took place in about 15 min. The hydrogel was allowed to age overnight and was then lyophilized to give the flaky white powder xerogel. The enzyme-containing xerogel was washed with water and buffers to remove any physically adsorbed enzyme. The xerogel produced in this manner seems to have a considerable degree of hydration.

According to Johnson and Whateley, enzyme entrapment occurs via the condensation of silicic acid sols to hydrogels having the three-dimensionally crosslinked networks composed of alternating silicon and oxygen atoms characteristic of silica. Initially, these networks result in only small but highly hydrated particles with little interparticle bonding. At a later stage, interparticle bonds become more important and these lead to gelation and formation of the hydrogel. The condensation reaction continues after the hydrogel has formed.

A method for immobilizing enzymes by entrapment within the microspace of fibers was announced recently.[375-377a] The enzyme molecules are entrapped in aqueous microdroplets within synthetic resin filaments which can be produced continuously by conventional wet-spinning techniques. A number of enzymes were immobilized and filaments of different chemical composition were described.

**Properties**

With this method of immobilization, no changes in the intrinsic properties of an enzyme are anticipated. Local microenvironmental effects such as those created by the nature of the carrier (e.g., its charge) or enzymatically generated ones created by the enzymic reaction itself can be expected. The charge of a carrier is an important consideration in trying to rationalize some of the effects observed. The charge of the water-insoluble matrix formed from polyacrylamide, starch, and Silastic resin is electrically neutral. On the other hand, the xerogel system formed from silicic acid sol is negatively charged at normal pH ranges (3 and above) due to the relatively easy ionizable silanol hydroxyl groups.

*Activity*

The activity of lattice-entrapped enzymes is critically dependent on the method of preparation. The relative activity of these water-insoluble conjugates is usually low but can go up to approximately 50 to 60% in the more favorable cases. Some examples of immobilized enzymes exhibiting low relative activities (with the percent activity given in parentheses) are trypsin (4 to 5.5),[355] (2),[358] (11),[378] α-chymotrypsin (4.5),[355] papain (3.4 to 6),[355] α-amylase (1.9),[355] β-amylase (6.6),[355] ribonuclease (4.6),[355] aldolase (4.2),[355] alcohol dehydrogenase (5),[378] lactate dehydrogenase (1),[378] steroid $\Delta^1$-dehydrogenase (7),[379] citrate synthase (12 to 15),[106] and glucose oxidase (15),[380] all immobilized in polyacrylamide gels. Relative activities in the range of 20 to 40% were reported for orsellinic acid decarboxylase (26 to 30),[361] aldolase (ca. 20),[360] phosphoglycerate mutase (up to 64),[381] and glucose isomerase (22 to 35),[366] likewise entrapped in polyacrylamide gels. The highest relative activity (ca. 60%) for lattice-entrapped enzyme conjugates reported to date was observed for phosphoglycerate mutase[381] entrapped in polyacrylamide and for trypsin and chymotrypsin[362] entrapped within Silastic resin. The relative activity of these immobilized enzymes (as with other water-insoluble enzyme conjugates) is dependent on the particular substrate employed for enzymic activity. This can be illustrated dramatically with trypsin immobilized in xerogel.[371] The relative esterase activity of this water-insoluble enzyme conjugate was 34% compared with soluble trypsin with BAEE as substrate.

However, this same trypsin derivative exhibited no detectable proteolytic activity toward casein as the substrate. Although diffusion and steric repulsion of the macromolecular species are certain to diminish the relative activity of this immobilized enzyme conjugate, some activity was still expected (see Chapter 3). The complete lack of activity in this instance was attributed by Johnson and Whateley to the unfavorable charged character of the substrate and immobilized enzyme (both being negatively charged) and the steric hindrance.

An interesting behavior of the relative activity with temperature of enolase immobilized in polyacrylamide was observed by Bernfeld and Bieber.[382] In their system, the relative activity of the insoluble form varied nonlinearly with temperature with a minimum occurring at 24°. At higher or lower temperatures, the ratio of the specific activity of the immobilized enzyme to the native enzyme increased.

*pH-Activity Behavior*

Although very few studies have been conducted to date on the pH-activity behavior of lattice-entrapped enzymes, the same observations have been made with these immobilized enzymes as with other enzyme conjugates. The pH-activity curves of the immobilized enzyme can be displaced toward either alkaline or acid pH values or need not be displaced at all. For example, trypsin[371] immobilized in xerogel exhibited the expected shift of its pH-activity curve toward more alkaline pH's. A shift of approximately 0.8 pH units was observed at an ionic strength of 0.2 with BAEE as substrate. No shift (or only a very negligible shift and well within experimental error) in the pH-activity curves was reported for several enzymes immobilized in polyacrylamide gel. For example, no shift was reported to occur with ribonuclease,[355] enolase,[382] or other enzymes.[355] This is the expected behavior due to the electrically neutral character of polyacrylamide. However, an anomalous pH-activity behavior has been observed for phosphoglycerate mutase[381] immobilized in polyacrylamide. An unexpected and as yet unexplained shift toward more acid pH's was observed for the lattice-entrapped enzyme. The pH-optimum of the water-insoluble, enzyme conjugate shifted to 6.5 (pH optimum of soluble enzyme, pH 7.3).

*Michaelis Constant*

A slightly higher apparent Michaelis constant was found for glucose oxidase[369] entrapped in Silastic resin (4 m*M*, compared with 2 m*M* for the soluble enzyme) and for urease[370] immobilized in polyacrylamide gel (5 m*M* and 4 m*M*, respectively). Similar $K'_m$ values were observed for immobilized phosphoglycerate mutase,[381] enolase,[382] and lactate dehydrogenase[358] immobilized in polyacrylamide. Decreased values of the apparent Michaelis constants were observed for cholinesterase[370] in starch gel ($K'_m$ of 0.16 m*M* and $K_m$ of 0.25 m*M* for immobilized and soluble enzymes, respectively) and for acetylcholinesterase[369] in silicone rubber (50 m*M* and 120 m*M* for immobilized and soluble enzyme, respectively). No explanation for the decreased $K'_m$ values was offered.

*Specificity*

Several publications have mentioned aspects of substrate and inhibitor specificity for entrapped enzymes. Brown et al.[362] noted that the calcium ion activation optima for soluble and polyacrylamide-immobilized apyrase were identical (1 m*M*) but that the extent of stimulation was different. High concentrations of $Ca^{+2}$ inhibited the entrapped enzyme. Pennington et al.[368] observed that acetylcholinesterase entrapped in silicone rubber was less affected by inhibitors than the freely soluble enzyme. Bernfeld et al.[381] observed similar substrate behavior for soluble and polyacrylamide-immobilized phosphoglycerate mutase. Pronounced substrate inhibition started to become noticeable at higher substrate concentrations for both enzymes. Polyacrylamide-immobilized enolase[382] exhibited different behavior toward magnesium ion inhibition. Both the soluble and immobilized enzymes required $Mg^{+2}$ for maximum activity (0.68 m*M*) but the entrapped enolase, in contrast to the soluble enzyme, was not inhibited by an excess of magnesium. Zinc ions inhibited both enzymes to about the same extent. At low magnesium ion concentrations, the polyacrylamide-enolase conjugate was somewhat less affected by $Zn^{+2}$ than was the soluble enolase.

*Stability*

Enhanced thermal stability (compared with the soluble enzyme) was observed for lactate dehydrogenase,[358] apyrase,[362] and trypsin[365] immobilized in polyacrylamide gel and for acetylcholinesterase[368] entrapped in Silastic resin. A higher temperature optimum was noted for enolase[382] immobilized in polyacrylamide gel. Similar thermal stability was noted for glucose oxidase[358] immobilized in polyacrylamide gel. A similar temperature optimum was noted for phosphoglycerate mutase[381] in polyacrylamide gel. The temperature optima for both the soluble and immobilized enzymes were between 40 to 45°. There was, however, a slight change in the shape of the activity-temperature curve of the immobilized enzyme compared with the curve of the soluble enzyme. Diminished stability has also been reported. Guilbault and Hrabankova[383] reported that the polyacrylamide-entrapped L-amino acid oxidase was less stable than the cellophane-entrapped enzyme. In this example, it is not clear whether the diminished stability was solely thermal.

Storage stabilities have been reported frequently. Ribonuclease A[355] immobilized in polyacrylamide retained 99% of its original activity after one month at 0 to 4°, lactate dehydrogenase[358] lost 10% activity per month over a period of three months, glucose oxidase[358] lost no activity during three months of storage at 0 to 4°, and orsellinic acid decarboxylase[361] lost 3% of its activity after 14 days at 20°. Storage stabilities of other enzymes immobilized in polyacrylamide gel have been reported for lactate dehydrogenase,[378] steroid $\Delta^1$-dehydrogenase,[379] citrate synthase,[106] and glucose-6-phosphate dehydrogenase.[107] Cholinesterase[370] immobilized in starch gel lost 2.5% of its original activity after 5 weeks of storage at ca. 4° and 32% after 20 weeks. At room temperature, 5% of its activity was lost in 5 weeks and 41.5% in 20 weeks. Trypsin[371] immobilized in silica gel lost less than 10% of its activity in 75 days stored at 4°. At room temperature, the storage stability was considerably reduced; after 70 days, only 27% of the original activity remained.

Stability of lattice-entrapped enzymes during continuous use was reported for cholinesterase[372] and for urease.[384]

Lattice-entrapped enzymes can be often lyophilized without serious inactivation.[358,371,372,379]

### Advantages and Disadvantages

Advantages of the lattice-entrapment method for immobilizing an enzyme include its overall experimental simplicity, the need for only small amounts of an enzyme in order to produce a water-insoluble enzyme conjugate, and the fact that it is a physical method. No chemical modification of the enzyme is expected and consequently no change in an enzyme's intrinsic properties is anticipated. The method also allows for a considerable choice of neutrally charged water-insoluble carriers. Perhaps most important, this method permits the preparation of water-insoluble enzyme derivatives of widely different physical forms. The often gelatinous nature of the enzyme derivative makes it easy to deposit an immobilized enzyme on either regular or highly irregular surfaces.

Disadvantages of the method also exist. Operationally, good immobilization (i.e., good mechanical properties and activity) is dependent on a delicate balance of experimental factors. The exact physical nature of the crosslinked polymers is often quite important for obtaining high activity. Chemical and thermal inactivation of enzymes during the gel formation can also take place. The formed lattice-entrapped enzyme often needs to be broken up or cut into suitable form. A considerable disadvantage of this method is leakage of an enzyme from within the crosslinked polymeric network. During formation of the polymer network and entrapment of an enzyme, differently sized microspaces (micropores) are created, the size distribution of which is determined largely by the relative degree of crosslinking. The larger the micropores, the greater will be the leakage. Although leakage, in principle and practice, can be reduced by reducing the size of these micropores, some initial leakage of enzyme molecules is certain to occur.[355,385] Severe initial leakage occurs especially with starch gel-immobilized enzymes.[370,372,373] Another major disadvantage of the method is the limitation to only small-sized substrates; lattice-entrapped enzymes show very little activity toward macromolecular substrates.

TABLE 10

**Lattice-Entrapped Enzyme Conjugates**

| Enzyme immobilized | Entrapment matrix[a] | Ref. |
|---|---|---|
| Alcohol dehydrogenase (1.1.1.1) | Polyacrylamide | 378 |
| Lactate dehydrogenase (1.1.1.27) | Polyacrylamide | 358, 378 |
| Glucose-6-phosphate dehydrogenase (1.1.1.49) | Polyacrylamide | 107 |
| Glucose oxidase (1.1.3.4) | Polyacrylamide | 358, 380, 386, 387 |
| | Starch | 372 |
| Glucose oxidase – peroxidase | Silastic resin | 369 |
| Steroid $\Delta^1$-dehydrogenase (1.3.-.-) | Polyacrylamide | 379 |
| Glutamate dehydrogenase (1.4.1.2) | Polyacrylamide | 358 |
| L-Amino acid oxidase (1.4.3.2) | Polyacrylamide | 383 |
| D-Amino acid oxidase (1.4.3.3) | Polyacrylamide | 388 |
| Amino acid oxidase (1.4.3.?) | Polyacrylamide | 358 |
| Catalase (1.11.1.6) | Polyacrylamide | 358 |
| | Silica gel (hydrogel) | 356 |
| Peroxidase (1.11.1.7) | Polyacrylamide | 389 |
| Hexokinase (2.7.1.1) | Polyacrylamide | 363 |
| Hexokinase – glucose-6-phosphate dehydrogenase | Polyacrylamide | 107 |
| Phosphofructokinase (2.7.1.11) | Polyacrylamide | 363 |
| Phosphoglycerate mutase (2.7.5.3) | Poly(*N,N'*-methylenebis-acrylamide) | 381 |
| Ribonuclease A (2.7.7.16) | Poly(*N,N'*-methylenebis-acrylamide) | 355 |
| Acetylcholinesterase (3.1.1.7) | Silastic resin | 368, 369 |

TABLE 10 (Continued)

**Lattice-Entrapped Enzyme Conjugates**

| Enzyme immobilized | Entrapment matrix[a] | Ref. |
|---|---|---|
| Cholinesterase (3.1.1.8) | Polyacrylamide | 364, 370 |
| | Silastic resin | 370 |
| | Starch | 370, 372-374 |
| Alkaline phosphatase (3.1.3.1) | Polyacrylamide | 359 |
| α-Amylase (3.2.1.1) | Poly(*N,N'*-methylenebis-acrylamide | 355 |
| β-Amylase (3.2.1.2) | Poly(*N,N'*-methylenebis-acrylamide) | 355 |
| Trypsin (3.4.4.4) | Polyacrylamide | 361, 365, 378 |
| | Poly(*N,N'*-methylenebis-acrylamide) | 355 |
| | Silastic resin | 362 |
| | Silica gel (xerogel) | 371 |
| α-Chymotrypsin (3.4.4.5) | Poly(*N,N'*-methylenebis-acrylamide) | 355 |
| | Silastic resin | 362 |
| Papain (3.4.4.10) | Poly(*N,N'*-methylenebis-acrylamide) | 355 |
| Asparaginase (3.5.1.1) | Polyacrylamide | 388 |
| Glutaminase (3.5.1.2) | Polyacrylamide | 390 |
| Urease (3.5.1.5) | Polyacrylamide | 370, 384, 385, 391-393 |
| | Silastic resin | 370 |
| | Silica gel (hydrogel) | 356 |
| | Starch | 370 |
| AMP deaminase (3.5.4.6) | Silica gel (hydrogel)[b] | 356 |
| Apyrase (3.6.1.5) | Polyacrylamide | 362 |
| | Silastic resin | 362 |
| Orsellinic acid decarboxy-ase (4.1.1.-) | Polyacrylamide | 361 |
| Aldolase (4.1.2.13) | Polyacrylamide | 363 |
| | Poly(*N,N'*-methylenebis-acrylamide) | 355, 360, 367 |

TABLE 10 (Continued)

**Lattice-Entrapped Enzyme Conjugates**

| Enzyme immobilized | Entrapment matrix[a] | Ref. |
|---|---|---|
| Citrate synthase (4.1.3.7) | Polyacrylamide | 106 |
| Enolase (4.2.1.11) | Poly(*N*,*N'*-methylenebis-acrylamide) | 382 |
| Glucose isomerase (5.3.1.-) | Polyacrylamide | 366 |
| Glucosephosphate isomerase (5.3.1.9) | Polyacrylamide | 363 |

[a]Polyacrylamide – mixture of acrylamide and *N*,*N'*-methylenebisacrylamide.
[b]No activity observed.

Chapter 7

# MICROENCAPSULATION

Enzymes can be immobilized within microcapsules that have either a permanent or nonpermanent semipermeable membrane. Permanent membranes are formed by interfacial polymerization or by coacervation of preformed polymers. Nonpermanent membranes or "liquid-surfactant membranes" are formed by the combination of appropriate surfactants, "additives," and hydrocarbons. Both of these interesting and potentially highly useful microcapsules are discussed here.

**Immobilization with Permanent Microcapsules**

The immobilization of enzymes by entrapping the molecules within permanent semipermeable microcapsules was first reported by Chang in the mid-'60's.[394-396] Since that time, various enzymes have been immobilized in microcapsules of different chemical composition (see Table 11). The principle of continuous operation using microencapsulated enzymes is based on the permselectivity of the membrane. The enzyme molecules, being larger than the mean pore diameter of the spherical membrane within which they are entrapped, cannot diffuse through the membrane into the external solution. On the other hand, substrate molecules whose size does not exceed the diameter of the pore can readily diffuse through the membrane and be transformed to product by the entrapped enzyme molecules. The product(s) of the reaction then diffuse(s) through the membrane to the exterior phase (see Figure 18). Only substrates and products of rather low molecular weights are applicable in this method. Typical microcapsules are shown in Figure 19.

Prior to Chang's first report, microencapsulation had been used for entrapping drugs, perfumes, detergents, dyes, adhesives, solvents, paints, chemicals, etc. A good general review of such microencapsulation was published by Herbig.[397] Usually, a slightly permeable or even nonpermeable membrane was employed for the encapsulation of these materials and sudden release of the encapsulated substance was dependent on the rupture of the membrane by pressure or heat or by the dissolution of the membrane itself. The success of Chang's method, however, depends on having a membrane as permeable as possible to the substrate and product but not to the enzyme.

*Preparation of Permanent Microcapsules*

Two general methods, based on coacervation and on interfacial polymerization, have been employed for the preparation of semipermeable microcapsules used for immobilizing enzymes.[394,396,398-400] Coacervation is the phenomenon of phase separation in polymer solutions, and the formation of a microcapsule is dependent on the lower solubility of the polymer at the interface of a microdroplet. Coacervation is a physical phenomenon. The interfacial polymerization method for producing a microcapsule is based on a chemical process – the synthesis of a water-insoluble copolymer at the interface of a microdroplet. One reactant is partially soluble in both the aqueous and organic phase and the other reactant (the second component of the copolymer) is soluble only in the organic phase. The

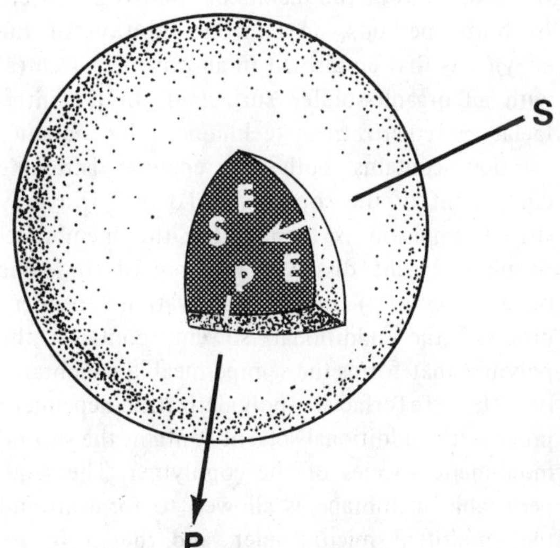


FIGURE 18. Schematic representation of a permanent microcapsule showing entrapped enzyme molecules (E). Continuous conversion is achieved by the diffusion of substrate (S) and product (P) molecules across the semipermeable membrane.

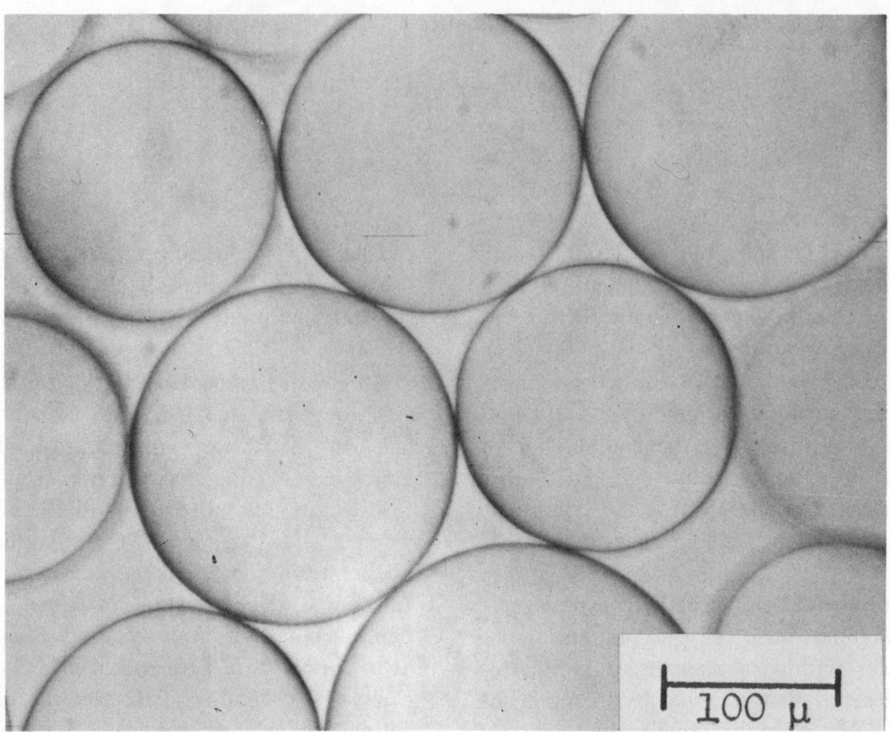


FIGURE 19. Photomicrograph of typical nylon-6,10 microcapsules containing entrapped soluble protein. Unpublished results, O. R. Zaborsky and J. Ogletree. Photographed by R. Sherwood.

partition coefficient of the water-soluble reactant between the aqueous and organic phase determines the properties of the membrane that is produced. In both methods, an aqueous solution of the enzyme is first emulsified in an organic solvent(s) with an organic-soluble surfactant. In the interfacial polymerization technique, the aqueous solution contains both the enzyme and one component of the copolymer. To the vigorously stirred emulsion is then added the membrane-forming reagent dissolved in more of the same organic solvent(s). In the coacervation-dependent process, the additional solvent contains the polymer that forms the semipermeable membrane; in the interfacial polymerization-dependent process, the additional solvent contains the second monomeric species of the copolymer. The semipermeable membrane is allowed to form around the emulsified microdroplet, and the microcapsules containing the entrapped enzyme are washed and transferred to a totally aqueous system with a water-soluble surfactant. In the coacervation-dependent process, "hardening" of the microcapsules is required before washing and transferring are possible. This final forming of the permanent membrane is usually achieved by addition of another water-immiscible organic solvent. Commonly employed organic-soluble and water-soluble surfactants are Span® 85 and Tween® 20, respectively.

Permanent microcapsules, produced by coacervation and employed for the immobilization of enzymes, have been made from collodion (cellulose nitrate),[394-396,398,400-404] polystyrene,[394,395] benzalkonium-heparin-collodion,[399,404] ethylcellulose,[405] and cellulose acetate butyrate.[255,256] Membrane formation has been achieved also with silicone type polymers (sesquiphenylsiloxane[405,406] and Silastic[401]) through a secondary emulsion technique. In this procedure, an initially produced water-oil emulsion is added to another aqueous phase in order to form permanent microcapsules.

Microcapsules formed by the interfacial polymerization method have so far been restricted to the well-known polyamide (nylon) system (see Equation 53). Although many different diamines and diacid

$$H_2N-(CH_2)_6-NH_2 \quad + \quad Cl-\overset{O}{\overset{\|}{C}}-(CH_2)_8-\overset{O}{\overset{\|}{C}}-Cl \xrightarrow[-HCl]{} \tag{53}$$

Hexamethylenediamine    Sebacoyl chloride

$$-NH-(CH_2)_6-NH-\overset{O}{\overset{\|}{C}}-(CH_2)_8-\overset{O}{\overset{\|}{C}}-NH-(CH_2)_6-NH-\overset{O}{\overset{\|}{C}}-(CH_2)_8-\overset{O}{\overset{\|}{C}}-NH-$$

Nylon – 6,10

halides can be used to prepare the amide copolymer, all studies reported on *enzyme* microencapsulation have used 1,6-diaminohexane (hexamethylenediamine) as the water-soluble diamine and 1,10-decanoyl chloride (sebacoyl chloride) as the organic-soluble diacid halide.[394–396,398,400–402,404,407–409] The organic solvent of choice for the preparation of nylon-6,10 microcapsules from hexamethylenediamine and sebacoyl chloride is a chloroform-cyclohexane mixture (1:4 v/v) containing usually 1% Span 85 (v/v). This particular ratio of chloroform to cyclohexane is used because the resulting mixture has the appropriate specific gravity and the correct composition for producing experimentally manageable microcapsules. The specific gravity of the organic solvent should be approximately 1, for if it is appreciably greater, the microcapsules tend to rise to the top of the solvent and are difficult to centrifuge. If the specific gravity is considerably less than 1, the microcapsules tend to sink to the bottom of the container and severe and irreversible aggregation can result. The specific gravity of the 1:4 (v/v) chloroform-cyclohexane mixture is 0.91.[396] Even more important than the favorable operational advantages gained by having the specific gravity approximately 1 is the fact that this solvent mixture has the appropriate partition coefficient for the diamine between aqueous and organic solvent phases. The nylon copolymer is formed on the organic side of the interface and the transfer of the water-soluble diamine to the organic phase is a prerequisite for the formation of these microcapsules.[411] The partition coefficient, K (defined as $C_{H_2O}/C_{solvent}$), of hexamethylenediamine in a water-chloroform mixture has a small value (0.70, measured at 25° at equilibrium with 2 moles of NaOH/mole of diamine in the aqueous phase) and hence if chloroform were used alone as the organic solvent, the microcapsules would have a thick membrane. Conversely, because of the large value of the partition coefficient of the diamine in a water-cyclohexane mixture (182, measured under the same conditions as above), the microcapsules produced by using this solvent would have a thin membrane. The same concentration levels of reactants are, of course, assumed in both systems. The solvent mixture of 1:4 chloroform-cyclohexane used and recommended by Chang[394,396] produces membranes of sufficient mechanical stability and strength.

Polyamide microcapsules are neutral. Negatively charged polyamide membranes can, however, be produced by the polymerization of 1,6-diaminohexane, 4,4′-diamino-2,2′-diphenyldisulfonic acid and sebacoyl chloride.[394,396,398] Nylon microcapsules, once formed, can also be made nonthrombogenic by coating the preformed capsules with benzalkonium-heparin-collodion solution or sequentially with heparin and benzalkonium solutions.[399] Microcapsules can be made even magnetic by introducing a slight amount of iron within the membrane wall during their formation.[401,408] These microcapsules can be readily removed from biological materials after in vivo studies and can be prevented from clogging the pores of filters in cartridge systems (extracorporeal shunts). Clogging of the pores is prevented, or at least retarded, by reversing the magnetic field.[401]

Polyamide semipermeable microcapsules can be formed also from just protein and the diacid halide without the addition of a low molecular weight diamine. Hemoglobin emulsified in a chloroform-cyclohexane mixture and then treated with sebacoyl chloride formed well-defined and stable microcapsules.[396]

Important experimental considerations for good and successful microencapsulation of enzymes have been reported by various investigators. Their general validity still remains to be thoroughly tested but anyone contemplating using microencapsulation as a means for immobilizing enzymes should be cognizant of them.

It is essential to have a concentrated solution of the enzyme in order that permanent membrane formation will occur.[394,396,398] For example, a 0.15 mg/ml solution of urease was needed for successful encapsulation of the enzyme within a polyamide membrane.[398] This high enzyme concentration can be circumvented by using a concentrated solution of another protein in addition to the enzyme to be immobilized. An erythrocyte hemolyzate, bovine serum albumin solution and a crude enzyme fraction have been used for this purpose. Boguslaski and Janik[409] reported that bovine serum solutions up to 50% were necessary in order to encapsulate successfully carbonic anhydrase within nylon microcapsules. A high protein concentration is evidently necessary in order to create a high osmotic pressure within the microdroplet so that membrane collapse cannot occur.

Although an organic-soluble surfactant, such as Span 85, is most helpful in producing stable and appropriately sized emulsified microdroplets, it is not necessary in every case. No emulsifying agent is needed if a sufficiently concentrated solution of the protein is used.

Emulsified microdroplets are formed normally by adding the aqueous enzyme solution to a vigorously stirred excess of the organic solvent. Magnetic stirring assemblies are used, and the size of the microcapsules so produced can be from 2 $\mu$ to several mm in diameter. Sparks et al.[408] described a syringe-based apparatus for making microcapsules (see Figure 20). In this apparatus, the aqueous enzyme-containing solution is forced through the small opening of a hypodermic needle, and a gas (at constant pressure and velocity) is fed through the larger concentric capillary. As a droplet begins to form, it extends into the gas stream and eventually detaches itself when its diameter is such that the drag force on its surface slightly exceeds the surface forces holding it to the orifice. The size of the microdroplets is determined by the velocity of the annular gas stream, and the size distribution is independent of the liquid flow rate. The lower limit of microdroplets produced by this technique was reported to be approximately 40 $\mu$ in diameter.

A "membrane-hardening" step is required in the phase-separation process for producing rigid and mechanically strong microcapsules. The procedure consists of transferring the initially formed microcapsules to another water-immiscible solvent system. For example, collodion-formed semipermeable microcapsules (in ether as the continuous phase) are hardened with *n*-butyl benzoate[394,396,398] containing Span 85 and cellulose acetate butyrate microcapsules (in a toluene-dichloromethane mixture) are hardened by addition of petroleum ether to the original solvent mixture.[255]

Transfer of the formed microcapsules to a totally aqueous phase is usually accomplished with Tween 20 – a step necessary for preventing aggregation of the microcapsules. A microcapsule washing apparatus was described by Sparks et al.[408]

### *Properties of Microcapsules*

Microcapsules prepared by either the interfacial polymerization or the coacervation techniques are nearly always spherical (see Figure 19). Occasionally, discoid,[396,398] cup-shaped, and even double[412] or multiple microcapsules can be produced. The microcapsules are rather flexible, and reversible crenation is possible.[398] Complete reversible crenation can occur in seconds to minutes if the porosity of the semipermeable membrane is sufficient enough to allow fast passage of solvent.

The mean diameter of microcapsules is dependent largely on the speed of mechanical emulsification, the concentration of the surfactant, and the viscosity of the organic liquid.[396,]

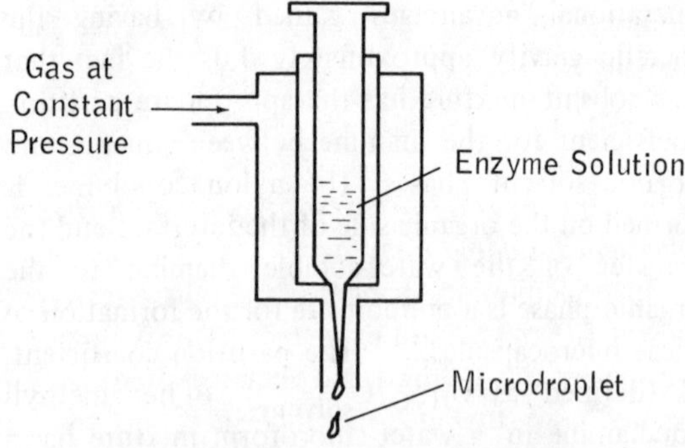


FIGURE 20. Syringe-based apparatus for producing microcapsules. (Redrawn from Sparks, R. E., Salemme, R. M., Meier, P. M., Litt, M. H., and Lindan, O., *Trans. Am. Soc. Artif. Intern, Organs,* 15, 353, 1969. With permission.[408])

[398] The mean diameter of microcapsules can be controlled and usually ranges from about 5 to 300 $\mu$. Larger or smaller microcapsules can be made, but the smaller ones ($< 5\ \mu$) are more difficult to prepare.

The thickness of the membrane and consequently its permeability is likewise influenced by a number of variables; the more important are the time of capsule formation, organic solvent composition, and the concentrations and the chemical nature of the components used to make the membrane. An appropriate value for the partition coefficient of the diamine is critical in controlling the thickness and permeability of polyamide microcapsules. Membrane thickness in 80 $\mu$ diameter nylon microcapsules made by the Chang procedure has been estimated to be about 0.2 $\mu$.[398]

The permeability characteristics of nylon and collodion microcapsules for various substances such as urea, creatine, glucose, sucrose, and uric acid were examined by Chang and Poznansky.[413] Permeability constants of these compounds for both types of membranes were found to be similar. Chang and Poznansky also estimated the equivalent pore radius of nylon microcapsules (270 $\mu$ in diameter and 200 Å thick) from values of the effective osmotic pressures of several electrolytes and found it to be approximately 18 Å.

Although the mechanical strength of microcapsules can vary considerably, they can be made to withstand normal handling operations. Microcapsules can be ruptured by pointed objects, compression, and by repeated freezing and thawing.

### *Properties of Permanent Membrane Microencapsulated Enzymes*

In principle, no changes in the inherent properties of an enzyme are expected upon its immobilization by this method. No chemical modification occurs, and the enzyme molecules are free in solution. Nevertheless, apparent changes in activity, Michaelis constant, pH-activity profile, etc. can be present because of diffusional effects of the membrane or enzymatically generated microenvironments.

#### *Activity*

The activity of a microencapsulated enzyme compared with its soluble counterpart has been reported to be anywhere from about 10 to essentially 100%. For example, the activity of microencapsulated carbonic anhydrase was given as "1/10 as fast as the enzyme in free solution, or 4/5 as fast as the erythrocyte-bound enzyme"[395] and also "essentially identical to that shown by the same concentration of enzyme in solution."[409] Comparative enzymic activities have been reported for urease[399,405] and catalase.[402,405] Even encapsulated homogenates[406] (those of zymase complex and rabbit muscle extract) showed activity in sequential reactions; both systems are known to involve at least 13 different enzymes for final product formation.

#### *Michaelis Constant*

$K'_m$ values have been reported only for carbonic anhydrase and L-asparaginase. Carbonic anhydrase[409] entrapped in nylon microcapsules with the aid of bovine serum albumin exhibited a 3 to 4 times greater value (36 to 59 m*M*) than the soluble enzyme (13 m*M*). Microencapsulated asparaginase[410] showed a $K'_m$ equal to that of the $K_m$ of the soluble enzyme.

#### *Specificity*

The specificity of microencapsulated enzymes is determined to some degree by the nature of the semipermeable membrane. Large substrates will not be able to diffuse through the membrane and will not be acted upon by the entrapped enzyme. Examples of specificity have been reported: microencapsulated trypsin[394] is still active toward dipeptide substrates but not toward proteins; carbonic anhydrase[394] is sensitive to acetazolamide inhibition; lipase[405] does not attack substrates such as olive oil and Tween 20.

#### *Stability*

Stability data have been given for several enzymes. The stability of microencapsulated carbonic anhydrase[394] was reported to be good. Crude urease[394] lost most of its activity in a day if a preparation of soya bean trypsin inhibitor was not also included, and asparaginase[410] retained 80% of its activity after 3 weeks of storage at 4°. Chang reported recently enhanced stability of microencapsulated catalase[246] at 4 and 37° relative to the soluble enzyme.

*Advantages and Disadvantages*

The microencapsulation of enzymes is a method of immobilization that provides an extremely large surface area for contact of substrate and catalyst, but all within a relatively small volume. For example, 20 $\mu$ diameter microcapsules have a surface area of 2,500 $cm^2/ml$,[401,407] and it has been calculated that it would be feasible to construct an artificial kidney (based on encapsulated urease and metabolite adsorbents) whose cylindrical dimensions would be only 10 cm in length and 2 cm in diameter.[407]

Microencapsulation also offers a "double specificity" due to both the enzyme and the semipermeable membrane. An initial selectivity can first be obtained by the appropriate choice of membrane. For example, if two low molecular weight substances of different ionic character were equally effective substrates for a particular enzyme, microencapsulation of the enzyme within a suitably charged semipermeable microcapsule would prevent entry of one of the substrates. The other substrate would diffuse readily into the microcapsule and be transformed by the enzyme.

Additional advantages are evident. The method allows the simultaneous immobilization of many enzymes in a single step. These enzymes can be water-soluble, previously immobilized by chemical attachment to water-insoluble functionalized polymers, natural membrane-bound enzymes, or multi-enzyme complexes.

Specific disadvantages of this method are the high protein concentration necessary for microcapsule formation, occasional inactivation of enzymes,[255,400,402] the possibility of enzyme incorporation into the membrane wall, and the restriction of substrates to low molecular weight substances. Leakage of enzyme from microcapsules does not appear to be a problem once good semipermeable microcapsules are prepared.

**Immobilization with Nonpermanent Microcapsules – Liquid-Surfactant Membrane Immobilization**

The immobilization of enzymes with nonpermanent microcapsules was disclosed recently by several investigators from the Esso Research and Engineering Company. The method is quite analogous to that described in the preceding section but has some potentially interesting and useful characteristics of its own. It is based on the "liquid-surfactant membrane" concept originated by Li[416] and subsequently developed by him[417-419] and others.[420]

The immobilization of an enzyme by this method involves encapsulating the protein solution within a semipermeable liquid-surfactant membrane.[421] The term "liquid-surfactant membrane" refers to a water-immiscible phase composed of surfactants, "additives," and a hydrocarbon solvent that contains emulsion-size aqueous droplets of various reagents or catalysts.[419] The size range of these emulsion-size aqueous droplets is approximately 1 to 100 $\mu$ in diameter. Usual membrane thickness is from 1 to 10 $\mu$ (see Figure 21).

*Preparation of Liquid-Surfactant Membrane Microcapsules*

The general procedure for immobilizing enzymes with liquid-surfactant membranes is similar to that used for microencapsulation with permanent membranes. The aqueous enzyme-containing solution is emulsified with a surfactant to form the liquid membrane encapsulated enzyme. Once prepared, the microcapsules are transferred to an aqueous solution to which then can be added the substrate. Substrates that can diffuse through the semipermeable liquid membrane are acted on by the entrapped and nondiffusible enzyme, and the product, once formed, can likewise diffuse through the membrane to the external aqueous phase. In this manner, a continuous enzymatic reaction is achieved. If desired, the enzyme can be immobilized in an aqueous membrane layer between two hydrocarbon-based interior and exterior phases.

The composition of the hydrocarbon membrane-forming solution can be varied extensively to give the appropriate membrane needed for a desired process. Usually the membrane is composed of a surfactant, "additives," and a high molecular weight paraffin. The surfactant most often used is Span 80. Additives are added to give the semipermeable liquid membrane both altered selectivity and physical stability. Polyethylene glycol, polyvinyl alcohol, and cellulose derivatives may be used as membrane-strengthening agents. The high molecular weight paraffin employed is an isoparaffin, Enjay® S100N, having an average molecular weight of about 390.

The simplicity and ease of the procedure for immobilizing enzymes may be illustrated by urease,[421] which was successfully immobilized by

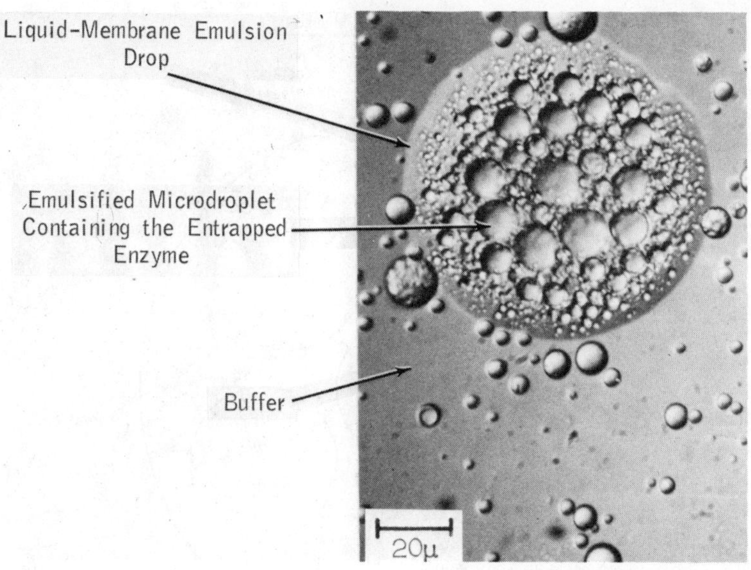


FIGURE 21. Interference contrast photomicrograph of liquid-surfactant membrane microcapsules containing phenolase. Enzyme system prepared by T. Hucal and S. W. May.

adding a 0.64 mg/ml solution of the enzyme to a vigorously agitated membrane-forming solution composed of 2% Span 80, 3% ENJ-3029 (a high molecular weight amine with an average molecular weight of 2,000), and 95% S100N.

*Properties of Liquid-Surfactant Membranes*

The nature of liquid-surfactant, semipermeable microcapsules is being studied and information about their structure and behavior is increasing. Figure 21 shows a representative example of hydrocarbon-based liquid surfactant membranes. Figure 22 gives a schematic representation of the presently held view of the structure of liquid membranes and how they are utilized for the immobilization of enzymes. Liquid membrane microcapsules are spherical and the size of the individual emulsion-size aqueous droplets within the assembly is usually 1 to 100 $\mu$ in diameter. The size of the microdroplets (the entire assembly) also varies and is typically in the range of 500 to 2,000 $\mu$ in diameter. As with any emulsion, the size of these liquid membrane microcapsules is dependent on a delicate balance and control of variables such as the speed of mechanical emulsification, the concentration of the emulsifying agent, the viscosity of the liquids, and the chemical nature of any and all reagents used. Membrane thickness is determined by the size of the reagent droplets and the ratio of reagent to organic phase; typically it is from 1 to 10 $\mu$.

The permeability and stability of liquid membrane microcapsules are dependent on the exact chemical nature of the membrane phase, the type and concentration of additives, and the temperature of the system. The permeability of substances across the liquid membrane microcapsules is solubility-dependent, and no diffusion of substrate or product will take place unless both have some solubility in the hydrocarbon-based liquid-surfactant membrane. This solubility dependency is in contrast to permanent microcapsules, where the diffusion of substrate or product is pore size-dependent. Additives can, however, be added to enhance the solubility of the permeate in a liquid membrane and thereby facilitate transfer and subsequent transformation. The stability of liquid membrane microcapsules can be varied considerably depending on the nature of the membrane forming solution, temperature of the system, etc. Liquid membrane microcapsules can be made that are stable for years, but they can also be broken by heat, centrifugation, or other emulsion-breaking techniques.

*Properties of Liquid-Surfactant Membrane Immobilized Enzymes*

To date, only urease,[421] phenolase,[422] and a

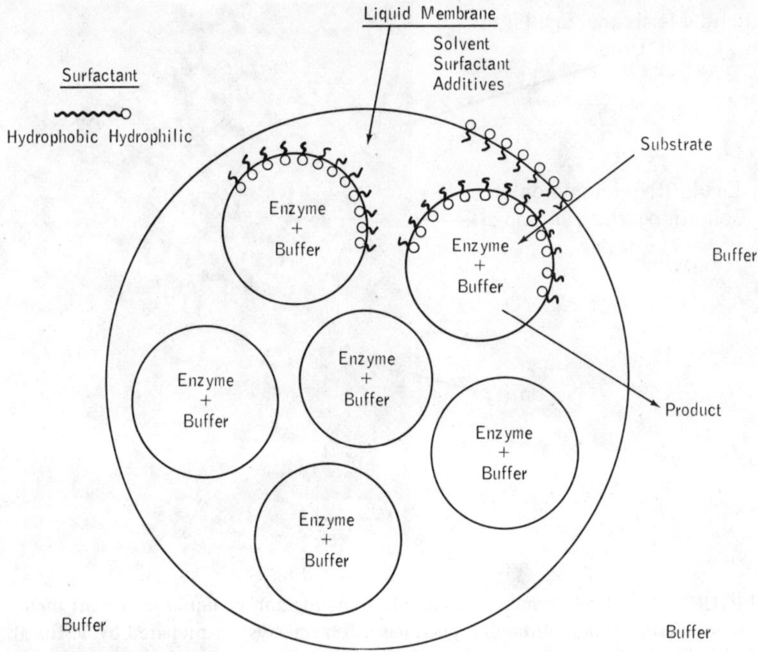


FIGURE 22. Schematic representation of the structure of liquid-surfactant membrane microcapsules, the immobilization of enzyme molecules, and the continuous catalytic conversion of substrate to product.

crude enzyme homogenate[423] (a nitrate reductase system) have been immobilized by liquid membrane microencapsulation. Of these, urease has been the enzyme that has been investigated more fully with regard to activity and possible leakage from the interior of the liquid membrane microcapsule. Urease, immobilized in the previously described hydrocarbon-based liquid-surfactant membrane, exhibited substantial catalytic activity, presented no leakage difficulties, and showed an apparent Michaelis constant approximately 50 times greater than the soluble enzyme.

*Advantages and Disadvantages*

The liquid membrane microencapsulation method for immobilizing enzymes has certain inherent advantages in addition to those mentioned for permanent microcapsules. It too provides for an extremely large surface area to volume ratio and for the simultaneous immobilization of many enzymes in one step. Specific advantages of the method are the nonpermanent, reversible, and completely nonchemical nature of the immobilization process. This immobilization method is dependent on a nonpermanent liquid-surfactant membrane and consequently it should be feasible to recover completely the immobilized enzyme by breaking the emulsified microdroplets by either chemical or physical means. A permanent *chemical* incorporation of the enzyme within the membrane wall is excluded.

Disadvantages of the method are possible "leakage" of the enzyme and the fact that transfer of substrates and products through the membrane is solubility-dependent.

TABLE 11

**Enzymes Immobilized with Permanent Semipermeable Microcapsules**

| Enzyme immobilized | Chemical composition of membrane[a] | Ref. |
|---|---|---|
| Urate oxidase (Uricase) (1.7.3.3) | Collodion | 400 |
| | Nylon | 400 |
| Catalase (1.11.1.6) | Collodion | 246, 400, 402, 404, 414 |
| | Not specifically stated (polystyrene, silicone, ethylcellulose) | 405 |
| Lipase (3.1.1.3) | Not specifically stated (polystyrene, silicone, ethylcellulose) | 405 |
| Trypsin (3.4.4.4) | Not specifically stated (nylon, collodion) | 394, 400 |
| L-Asparaginase (3.5.1.1) | Collodion | 246, 404, 414 |
| | Nylon | 246, 404, 410, 414 |
| Urease (3.5.1.5) | Collodion | 398 |
| | Nylon | 395, 398, 401, 407, 408 |
| | Benzalkonium-heparin-collodion | 399 |
| | Not specifically stated (nylon, collodion) | 246, 394, 400, 403, 404, 413-415 |
| | Not specifically stated (polystyrene, silicone, ethylcellulose) | 405 |
| Urease-silica-glutaraldehyde | Cellulose acetate butyrate | 255, 256 |
| Carbonic anhydrase (4.2.1.1) (sometimes as erythrocyte hemolyzate) | Collodion | 398, 400, 404, 414 |
| | Nylon | 394, 395, 398, 400, 404, 409, 414 |
| | Silicone | 401, 405 |
| | Not specifically stated (polystyrene, ethylcellulose) | 405 |
| Rabbit muscle extract (glucose to lactate) | Silicone | 406 |
| Zymase complex, yeast (glucose to ethanol) | Silicone | 406 |

[a]Chemical composition of microcapsules given in parentheses are the most likely compositions used for immobilizing the enzyme.

Chapter 8

# CONTAINMENT WITHIN SEMIPERMEABLE MEMBRANE DEVICES

Various devices that contain a semipermeable membrane have been used for the immobilization of enzymes. These are available commercially and are either of a cell form, employing a flat disk-like membrane, or of a beaker, T-tube, and shell-and-tube (cartridge) configuration, employing bundles of hollow fibers whose walls are the membranes. The devices can be considered simply as containers for the localization of an enzyme in much the same way as a beaker. Yet, because of the presence of a semipermeable membrane, they permit a continuous operational mode. Immobilization consists of putting the aqueous protein solution into the proper membrane-containing cavity of the apparatus. The successful operation of such an enzyme reactor is dependent on the permeability characteristics of the membrane employed. Ideally, the membrane should retain the enzyme completely but allow free passage of the product.

Two general membrane-dependent processes that have been used to achieve a separation of the product from the enzyme are ultrafiltration and dialysis. In ultrafiltration, a solution is filtered under a pressure gradient through a semipermeable membrane resulting in at least a partial separation of solute molecules from the solvent or from the solvent and other solute molecules. Ultrafiltration, reverse osmosis, and ordinary filtration differ superficially only in the size scale of the particles that are separated. Reverse osmosis is commonly used to describe the pressure-mediated, membrane-dependent separation of solutes whose molecular dimensions are within one order of magnitude of those of the solvent. Ultrafiltration, on the other hand, is used commonly to describe pressure-mediated, membrane-dependent separations of solutes having molecular dimensions at least 10 times greater than those of the solvent but less than approximately 5,000Å. Dialysis is also a membrane-dependent separation process, but it is not pressure mediated. In dialysis, the components transfer across the semipermeable membrane because of a concentration difference between the two liquid phases which, separated by the membrane, are at equal hydrostatic pressures.

Some of the devices can function either by dialytic or ultrafiltration processes; others can function only by the process of ultrafiltration or dialysis. The heart of the continuous operational mode of these devices is the semipermeable membrane, and a brief and general description of the nature and mechanism of transport through membranes follows.

## Nature of Semipermeable Membranes

Excellent reviews on ultrafiltration membranes[424,425] and on membrane permeation theory[426] were published recently. Van Oss[424] presented an excellent description of the various types of membranes used for ultrafiltration purposes. The history, chemical and physical nature, experimental procedures employed, and both advantages and disadvantages of each kind of membrane were covered. Van Oss arranged semipermeable ultrafiltration membranes into two general classes – homogeneous and anisotropic. Homogeneous or isotropic membranes have a uniform structure throughout the entire wall; the membrane has the same structure on its surface as it has in its interior. Anisotropic membranes, on the other hand, have a heterogeneous structure. This means that one side of the membrane has a smooth surface or "skin," while the other side has a rough surface and is more porous in nature. Anisotropic membranes consist of an extremely thin (0.1 to 10$\mu$) layer of homogeneous polymer supported upon a much thicker (20$\mu$ to 1 mm) layer of microporous sponge.

Michaels[425] likewise gave a description of the various types of membranes used for ultrafiltration purposes but proposed another classification system. He discussed the ideal characteristics of a good membrane, theoretical aspects of transport kinetics for "diffusive" and "microporous" membranes, "concentration polarization," and laboratory and industrial applications of ultrafiltration. In his view, a good ultrafiltration membrane should have a high hydraulic permeability to solvent, sharp "retention-cutoff" characteristics, good mechanical durability, good chemical and thermal stability, high flow stability, high fouling resistance, and excellent manufacturing reproducibility of flow and retention characteristics.

Michaels's classification of ultrafiltration membranes consists of two general types – the "microporous" and "diffusive" – both of which are subdivided into isotropic and anisotropic.

Microporous ultrafilters can be regarded as "classical" type filters because they have a rigid, highly voided structure containing interconnected, extremely small, random pores of approximately 500 to 5,000 Å in size. Solvent flow is essentially "viscous" in nature and flow rate is proportional to the pressure difference. Dissolved solutes whose molecular dimensions are smaller than the smallest pores pass through the membrane, but larger sized molecules become either entrapped or stacked on the surface of the membrane. The "cutoff level" of this type of membrane depends upon its mean pore size, and the "sharpness of cutoff" depends on the breadth of the pore size distribution. Plugging of microporous filters can be a severe problem and especially manifests itself with solutes whose dimensions lie in the lower third of the pore size distribution because these have the least difficulty entering the structure but the greatest likelihood of lodging in pore constrictions. Pore fouling can cause a dramatic reduction in the permeability of the solvent and a change in the selectivity of a membrane. A microporous ultrafilter whose mean pore size is far below that of the dimensions of the solute is recommended in order to prevent, or at least to diminish, solute plugging.

Diffusive ultrafilters can be regarded as "nonclassical" type filters. In these, both solute and solvent migration occur via random thermal (Brownian) movements of molecules between the chain segments of the polymer network. Solute and solvent molecules are transported by molecular diffusion under the action of a concentration or activity gradient. In general, a more highly expanded gel matrix has a relatively higher solute and solvent permeability than a correspondingly tighter matrix. The higher permeability is caused by the greater flexibility of the polymer chains. A diffusive ultrafilter contains no "pores" in the conventional or classical sense, and they are not plugged by retained solute because the concentration of any solute within the membrane is low and time-independent. There is no decrease of solvent permeability in these filters with time at constant pressure.

Presently available isotropic diffusive ultrafilters have characteristically low solvent permeabilities due to the rather thick semipermeable membrane, and these filters have, therefore, not been used often. In order to achieve a reasonable flow rate (100 ml water/ft$^2$/hr at 100 psi) in a diffusive type ultrafilter, an effective membrane thickness in the micron or submicron range is needed. Although it is not possible at the present time to prepare such a thin and perfect polymeric membrane, it is possible to prepare a diffusive type ultrafilter which has the sufficient polymeric membrane surface. These are the anisotropic diffusive ultrafilters. An electron micrograph of the cross section of such a membrane is given in Figure 23. Anisotropic diffusive membranes are used with the thin film or "skin" side exposed toward the high pressure solution. They may be supported on the spongy side with inert porous material to provide greater strength and durability. Filtration characteristics of these anisotropic ultrafilters are dependent largely on the nature of the thin surface layer.

### Immobilization with Ultrafiltration Cells

The use of ultrafiltration cells for the localization of an enzyme in order to carry out a continuous reaction has been reported by several investigators since the late 60's.[90,91,427–433] A clear distinction must be made between the conventional use of these ultrafiltration cells and their use as continuous enzyme reactors. Ultrafiltration cells had been used often for either concentration or purification of enzyme solutions, and the enzyme of interest was indeed localized in these instances. However, these modes of operation were not aimed at achieving a catalytic conversion with the aid of a catalyst that could be reused and recovered at will.

The several commercially available laboratory-sized ultrafiltration units range considerably in size, configurational design, type of fluid flow over the semipermeable membrane, etc. At the moment, the Amicon Corporation offers the greatest choice of the basic ultrafiltration cells, membranes, and needed auxiliary equipment. It is, therefore, not surprising to discover that almost exclusively Amicon systems have been used for enzyme immobilization. Amicon manufactures unstirred, magnetically stirred, and single or recirculating thin-channel (laminar flow) cells. The most commonly employed ultrafiltration cells for

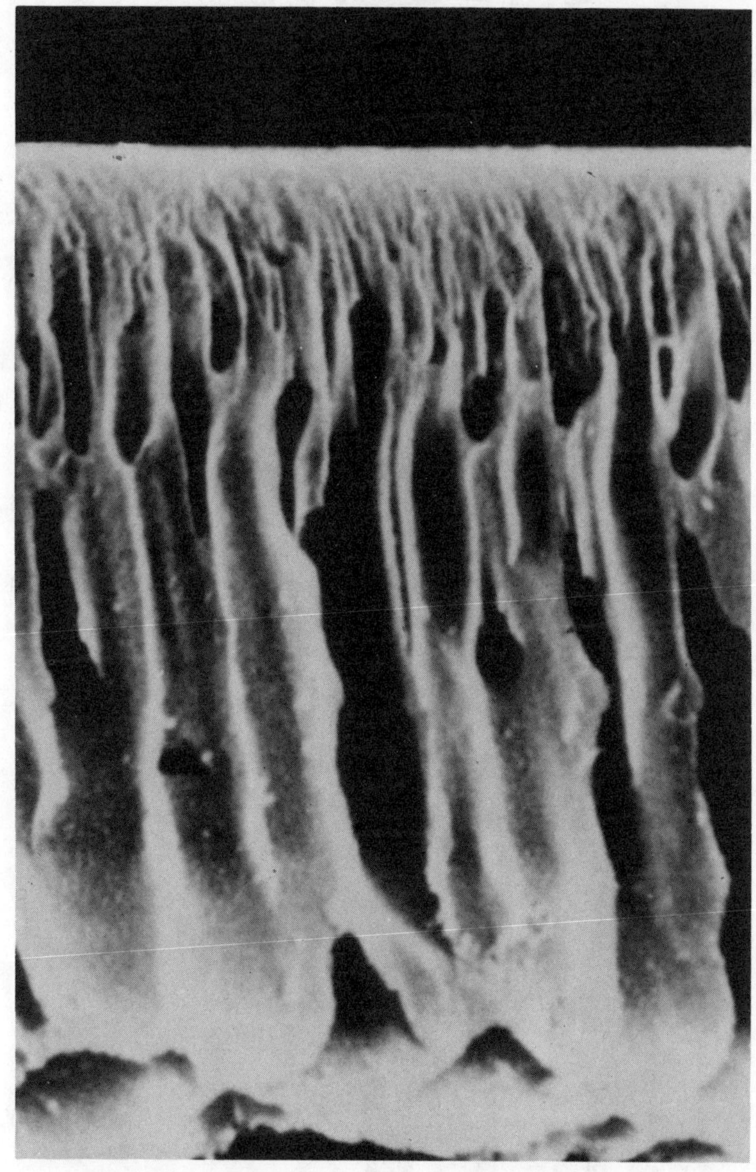

FIGURE 23. Photomicrograph of a typical Amicon Diaflo membrane (ca. 1000 X). Ultrafiltration "skin" on top of the anisotropic membrane is not visible at this magnification. (Courtesy of the Amicon Corporation, Lexington, Mass.)

immobilizing enzymes have been the magnetically stirred cells shown in Figures 24 and 25. A schematic representation of the operation of these cells is given in Figure 26. A recirculating thin-channel cell is shown in Figure 27. Thin-channel cells should be exceedingly useful in certain operations as continuous enzyme reactors because of their high throughput of solvent and solute and efficient operation even at the high protein concentrations. Somewhat similar ultrafiltration equipment can be obtained from the Abcor, Biomed, Chemapec, Gelman, Millipore, Sartorius, and Schleicher and Schuell companies. Because of the rather high activity and fast developments in this area, individuals contemplating using this equipment are advised to contact the companies for the latest developments and available apparatus.

The experimental procedure for the continuous catalytic conversion of a substrate and the separation of the product from the enzyme using a stirred ultrafiltration cell is illustrated in Figure

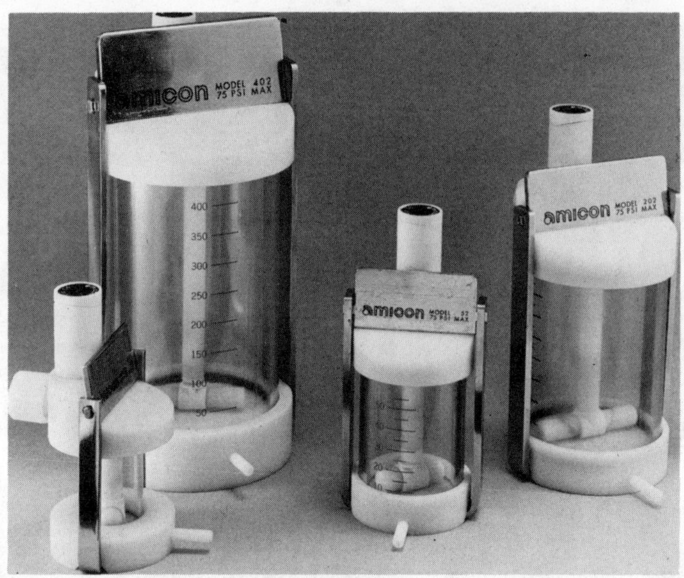



FIGURE 24. Conventional magnetically stirred Amicon ultrafiltration cells. (Courtesy of the Amicon Corporation, Lexington, Mass.)

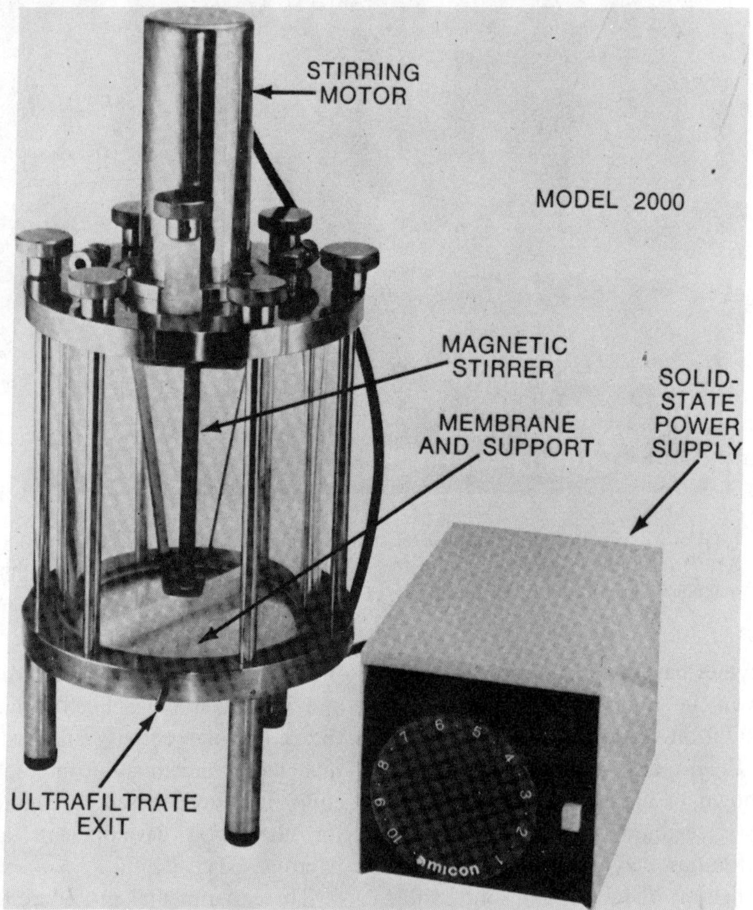



FIGURE 25. Amicon high-flow ultrafiltration cell with motor-driven, magnetically coupled stirring assembly. (Courtesy of the Amicon Corporation, Lexington, Mass.)

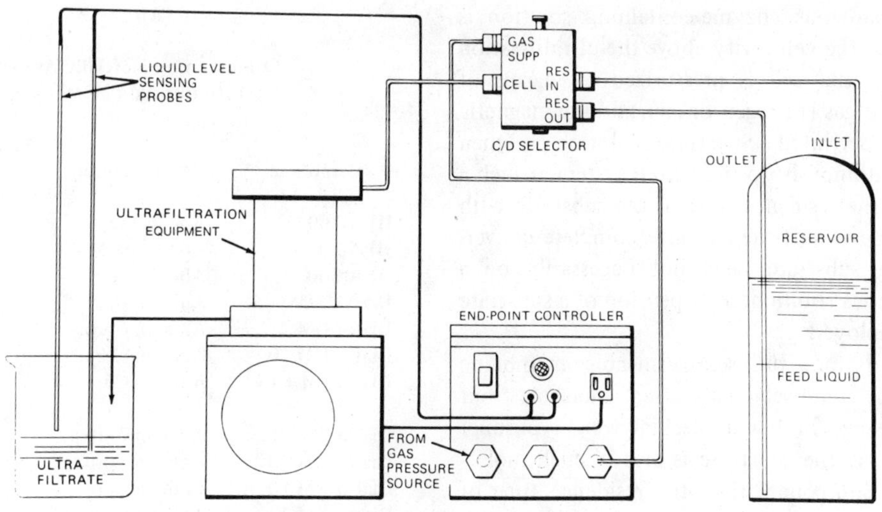


FIGURE 26. Typical ultrafiltration system for use in concentration or diafiltration or as a continuous enzyme reactor. (Courtesy of the Amicon Corporation, Lexington, Mass.)

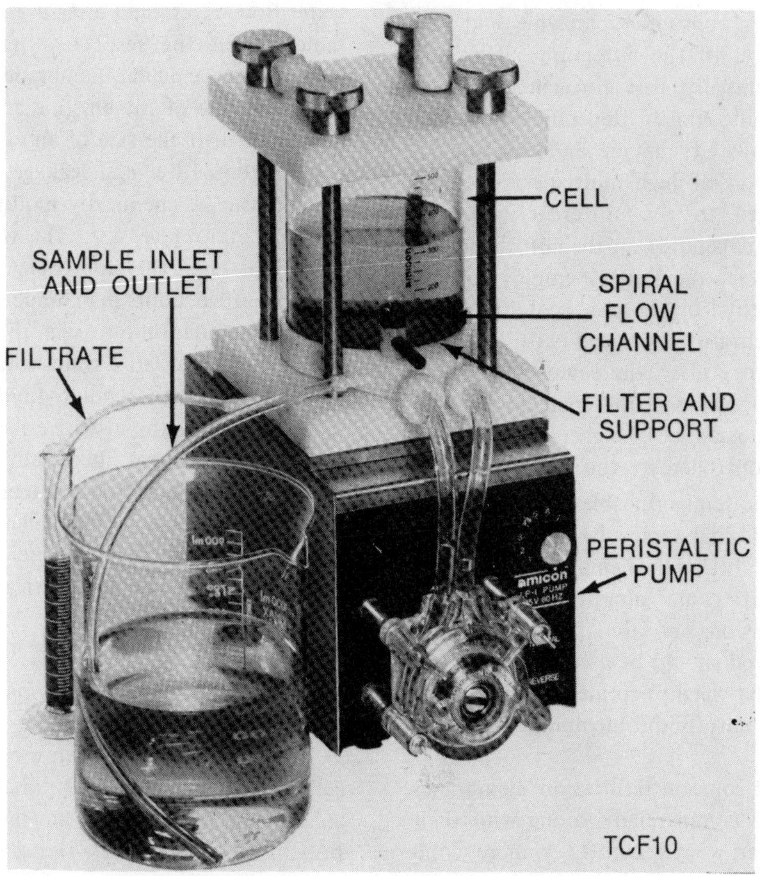


FIGURE 27. Recirculating, thin-channel ultrafiltration cell. (Courtesy of the Amicon Corporation, Lexington, Mass.)

26. The aqueous enzyme-containing solution is placed into the cell cavity above the ultrafiltration membrane, the cell is pressurized by means of compressed gas (nitrogen or air), and the magnetic stirrer is activated. Substrate solution is then added continuously to the stirred system at such a rate that the residence time of the substrate with the enzyme is sufficient to give complete conversion. The substrate need not necessarily be a homogeneous solution; a suspension of a substrate can be employed.

In operation, the semipermeable membrane retains the relatively large-sized molecules but allows passage of low molecular weight product molecules. If the substrate is also a small molecular-sized compound, then the residence time of the substrate in contact with the enzyme must be carefully controlled if 100% conversion is to be achieved. With high molecular weight polymeric substrates (such as starch or proteins) that are to be degraded to their constitutive components, this problem does not exist. The semipermeable membrane retains the appropriate enzyme and substrate molecules until the latter are sufficiently degraded to eventually pass through the filter. Within some limits, mostly depending upon the molecular size of the enzyme and its catalytic nature, degradation of high molecular substrates can be controlled to give different molecular weight products depending on the cutoff limit of the semipermeable membrane employed. For example, if the hydrolysis of a protein is conducted in an ultrafiltration cell with a rather nonspecific endopeptidase, the degree of degradation (i.e., the proportion of low molecular weight fragments to substantially higher weight fragments) can be controlled by the selection of the cutoff limit of the semipermeable membrane. If a large proportion of high molecular weight peptides is desired, an ultrafiltration membrane with a high cutoff limit that could almost approach the molecular dimensions of the particular endopeptidase employed should be used. Conversely, if only low molecular weight peptides or amino acids are desired, a low cutoff limit membrane should be used.

Table 12 lists some ultrafiltration membranes that are available commercially, along with their nominal molecular weight cutoff. A more complete description of a specific membrane's preparation, solvent and buffer compatibility, temperature stability, solute retention characteristics, water-flow rates, and pressure limits can be obtained from the respective manufacturers. The choice of a particular membrane is dictated by the molecular size of the enzyme and the products to be obtained. If the size of an enzyme is too small for a certain filter and leakage would occur, the enzyme can be chemically modified to increase its effective molecular size. The work of O'Neill et al.[91] illustrates this approach. In order to use a higher cutoff limit membrane in the $\alpha$-chymotrypsin degradation of casein, the enzyme was first covalently linked to a water-soluble dextran (mol wt ca. $2x10^6$) with 2-amino-4,6-dichloro-*s*-triazine. In this manner, the effective molecular weight of $\alpha$-chymotrypsin was increased so that a higher cutoff limit ultrafiltration membrane could be used (Amicon XM 100). Similarly, $\alpha$-amylase was covalently bonded to water-soluble amino-*s*-triazinyl derivative of dextran 2000, DEAE-dextran, and CM-cellulose.[90]

TABLE 12

**Common Commercially Available Ultrafiltration Membranes**

| Membrane | Manufacturer | Nominal molecular weight cutoff |
|---|---|---|
| HFA 100 | Abcor, Inc. | 10,000 |
| HFA 200 | Abcor, Inc. | 20,000 |
| HFA 300 | Abcor, Inc. | 70,000 |
| Diaflo® UM 05 | Amicon Corp. | 500 |
| Diaflo UM 2 | Amicon Corp. | 1,000 |
| Diaflo UM 10 | Amicon Corp. | 10,000 |
| Diaflo PM 10 | Amicon Corp. | 10,000 |
| Diaflo PM 30 | Amicon Corp. | 30,000 |
| Diaflo XM 50 | Amicon Corp. | 50,000 |
| Diaflo XM 100A | Amicon Corp. | 100,000 |
| Diaflo XM 300 | Amicon Corp. | 300,000 |
| PSAC Pellicon | Millipore Corp. | 1,000 |
| PSED Pellicon | Millipore Corp. | 25,000 |
| PSJM Pellicon | Millipore Corp. | 100,000 |

### *Properties of Ultrafiltration Cell-Contained Enzymes*

Enzymes that have been used in ultrafiltration cells are given in Table 13. Inherent properties of an enzyme immobilized in this manner are, in principle, the same as those exhibited by the enzyme in any other type of vessel. The enzyme is not chemically modified, and it can be in intimate contact with the substrate at all times in the well-mixed macroscopic environment. Microenvironmental effects such as those exhibited by enzymes in polyacrylamide gel immobilization should not be present, for diffusion of substrate and product are not a problem. Therefore, parameters such as activity, pH-optimum, catalytic rate constant, $K_m$, etc. should not be altered from the values obtained under identical conditions in other vessels. However, one important characteristic – the stability – of an enzyme could be altered to some degree. Inactivation of enzymes could occur by too vigorous agitation, adsorption of the enzyme on the membrane, or by high shear forces. Long-term stability studies of enzyme reactors have been reported for glucoamylase (120 hr at 40°,[433] 70 hr at 40°[429]), α-chymotrypsin covalently linked to dextran (2 weeks at 20°[91]), cellulase (243 hr at 50°[431]), α-amylase (12 days at 40°[429]), and α-amylase attached to CM-cellulose (70 hr at 70°[90]). Substantial activity was retained by the immobilized enzyme in every case, but these enzymes are rather stable macromolecules.

The study of immobilized α-amylase and glucoamylase by Butterworth et al.[429] merits further discussion. Although *retained* α-amylase and glucoamylase lost little activity upon containment in ultrafiltration cells, leakage of both enzymes occurred to a considerable extent through an Amicon PM 10 filter. The leakage loss for α-amylase (mol wt ca. 50,000) through the PM 10 filter (mol wt cutoff ca. 10,000) was about 35% during the first two volume replacements, after which essentially complete retention was achieved. This result seems somewhat perplexing because the molecular weight of the enzyme is substantially greater than the reported nominal molecular weight cutoff of the membrane. The authors attributed this initial leakage of α-amylase to a low initial rejection efficiency of the membrane which, in turn, was ascribed to both the compact globular shape of the enzyme and to the difference between the convective flux of the protein molecules toward the membrane and their ability to diffuse back into the bulk solution. This difference resulted in an accumulation of enzyme molecules at the membrane surface – a phenomenon called "concentration polarization." Consequently, because of the greater concentration of enzyme molecules at the membrane surface and the fact that transport behavior of a membrane is governed by its boundary environment, a portion of this highly concentrated layer of enzyme molecules is passed through the membrane. Then after several volume replacements occur, the formed enzyme layer or hydrogel acts as a secondary membrane which then controls the overall retention characteristics of the membrane, and a steady rate rejection of the enzyme is achieved. Similar initial membrane leakage was observed with glucoamylase[429] and an Amicon PM-10 membrane. Surprisingly, other investigators using the same enzymes and similar membranes have not reported this difficulty.

*Advantages and Disadvantages*

There are several advantages of immobilizing enzymes in ultrafiltration cells for conducting continuous enzymatic conversions. The method is exceedingly simple; immobilization is achieved by putting a solution of the enzyme into the ultrafiltration cell cavity. After substrate conversion is complete, the enzyme can be withdrawn and used again.

No chemical modification of the enzyme is required for its eventual immobilization. In fact, no "wet chemistry" is involved at all in the immobilization procedure because the ultrafiltration cells and membranes are available commercially. An immobilized enzyme will exhibit the same characteristics as it would in a plain beaker, and in principle, no irreversible modification or alteration takes place.

The method is especially suited for immobilizing soluble enzymes that act on high molecular weight water-soluble or water-insoluble substrates. The use of a soluble enzyme permits the necessary intimate contact of enzyme with substrate in order to achieve an efficient conversion of high molecular weight substrates. Water-insoluble, polymer-bound enzymes have characteristically lower catalytic efficiencies toward high than low molecular weight substrates because of the unavoidable steric repulsion of the polymer backbone and substrate.

The degree of product distribution from enzymatic degradation of high molecular weight substrates can be controlled to a fair extent by the cutoff limit of the membranes. A high cutoff limit membrane will give a higher proportion of the larger-sized fragments. Conversely, a low cutoff limit membrane will give essentially the constitutive monomeric units. The membrane retains the polymeric substrate molecules in contact with the enzyme until the product molecules are sufficiently small to pass through the membrane. In addition to a purely size-dependent separation, an appropriately charged membrane could be used to achieve an even further selective separation based on charge difference of low or high molecular weight substrates.

The method allows also for the simultaneous immobilization of many enzymes. This can be especially beneficial in dealing with labile multi-subunit enzymes or with several enzymes catalyzing a series of consecutive transformations.

General disadvantages of the ultrafiltration method for immobilizing enzymes are the possibility of enzyme inactivation due to vigorous agitation, adsorption, or high shear forces, the inherent limitations of the equipment which sometimes impose controls in pressure and filter selection, the possibility of enzyme leakage, and the precise control of residence time needed for complete or controlled conversions of low molecular weight substrates.

**Immobilization with Hollow Fiber Devices**

Semipermeable, hollow-bore fibers are being used increasingly and are being suggested for purposes such as water purification, food processing, separation of gases, blood oxygenation, and artificial kidney units. Companies that produce hollow fiber-containing devices of various designs and sizes are Abcor, Amicon, Dow, DuPont, and Monsanto. At the present time, only the Amicon Corporation and Dow Chemical Company are manufacturing laboratory-sized hollow fiber devices that can be used for ultrafiltration or dialytic processes. Dow hollow fiber devices are marketed by Bio-Rad Laboratories.

Some of the commercial units available from Amicon and Dow are shown in Figures 28 to 31. These units come in various shapes, sizes, and hollow fiber capacities for either ultrafiltration or dialytic modes. All units contain "bundles" of hollow fibers, an individual fiber of which is usually about 200 $\mu$ in internal diameter and approximately 25 $\mu$ thick.

Dow semipermeable hollow fibers are made of cellulose acetate or cellulose. They are homogeneous (isotropic) membranes and permit solute and solvent permeation from either the inside or outside wall of the fiber. Dow hollow fiber units can be employed for dialytic or ultrafiltration processes.

Amicon hollow fibers, on the other hand, are composed of the same materials as the PM 10 and

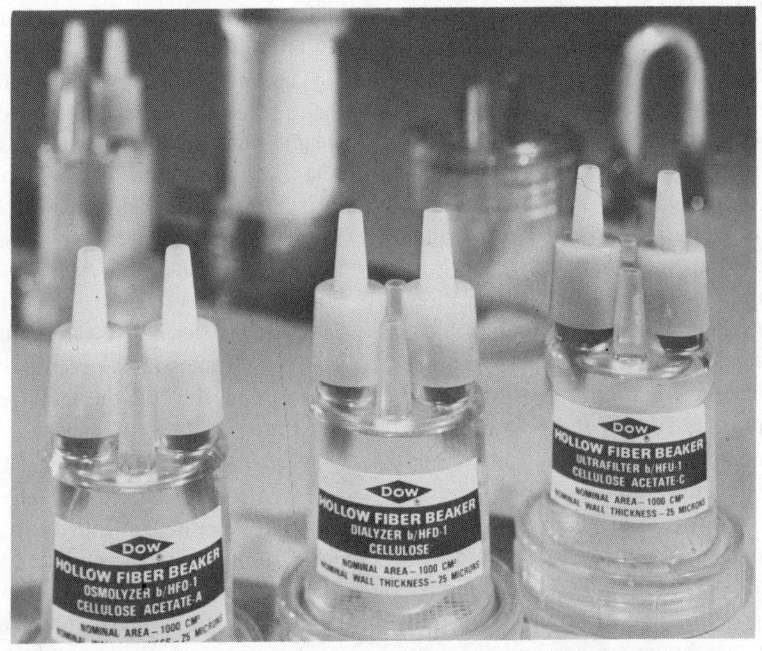


FIGURE 28. Dow hollow fiber devices. Beaker osmolyzer, dialyzer, and ultrafilter are seen in foreground. Other hollow fiber systems are also shown. (Courtesy of the Dow Chemical Company, Midland, Mich.) (From Form No. 175-1187-71, 1971.)

FIGURE 29. Dow hollow fiber beaker device shown in continuous operational mode. (Courtesy of the Dow Chemical Company, Midland, Mich.) (From Form No. 175-1187-71, 1971.)

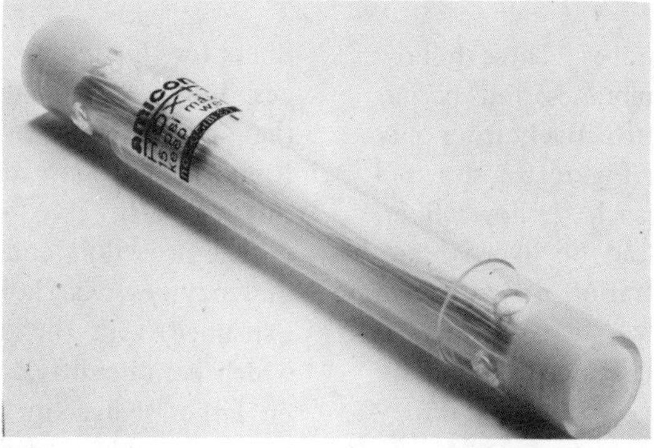


FIGURE 30. Amicon hollow fiber cartridge. (Courtesy of the Amicon Corporation, Lexington, Mass.)

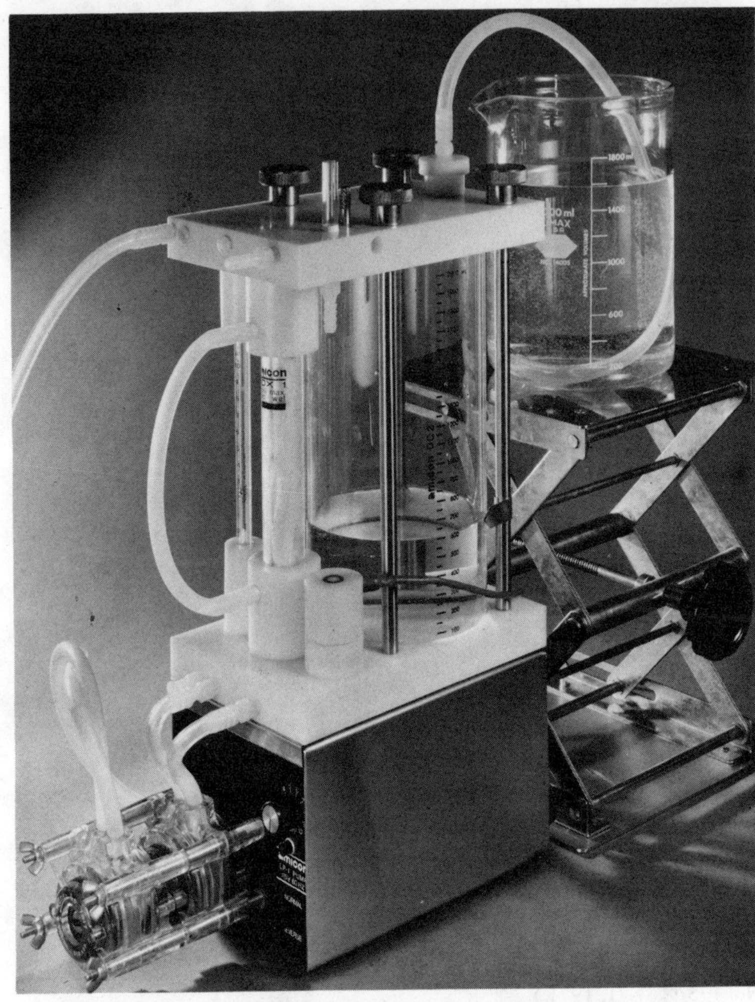

FIGURE 31. Amicon hollow fiber dialyzer/concentrator. (Courtesy of the Amicon Corporation, Lexington, Mass.)

XM 50 ultrafiltration membranes. These hollow fibers have anisotropic membranes and permit solute or solvent to permeate effectively from only one side of the fiber wall. Figure 32 shows a photomicrograph of one such hollow fiber. Amicon hollow fiber units can be likewise employed in dialysis or ultrafiltration processes. All hollow fibers have, of course, two distinct advantages over sheet membrane systems; they have a high ratio of membrane area to volume, and they are self-supporting.

Rony[434] first advanced the use of hollow fibers for enzyme immobilization in a theoretical paper on diffusion control within hollow fibers. He also compared this immobilization method with that of permanent membrane microencapsulation and discussed the advantages of hollow fibers for this purpose. Although no experimental description was included to verify his claims at that time, except for a note added in proof, Rony[435] has since demonstrated the validity of this approach.

The procedure and principle of immobilization of enzymes via hollow fiber devices can be explained with the use of a beaker type device which permits different modes of operation based on either dialysis or ultrafiltration. Some of the more favorable modes of operation for immobilizing an enzyme are shown in Figure 33. The arrows from the letters designating the substrate (S) and product (P) show the direction of flow. The enzyme can be contained either on the inside or the outside of the semipermeable hollow fibers. In both modes, A and B, the substrate is added to the

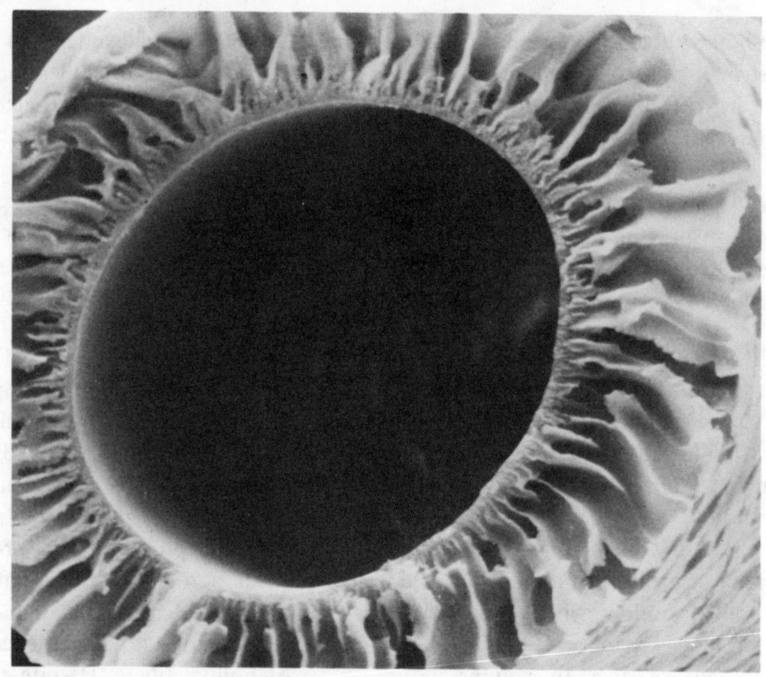

FIGURE 32. Photomicrograph of typical Amicon hollow fiber (300 X). (Courtesy of the Amicon Corporation, Lexington, Mass.)

enzyme-containing solution in a continuous fashion, and the product is continuously removed from the other phase once it has permeated across the membrane. As with the flat, disk-like ultrafiltration membranes, the success of the continuous enzyme reactor is dependent on the characteristics of the membrane. The membrane must be able to retain the enzyme completely, have some preferential retention for the substrate, and allow free transfer of the product. The use of low molecular weight substrates requires a careful control of the residence time in order to achieve high conversions. High molecular weight polymeric substrates would seemingly present no difficulties as is the case with the sheet-like ultrafiltration membrane systems. The interesting analogy between immobilizing enzymes within hollow fibers and permanent microcapsules, discussed by Rony, is illustrated by mode C in Figure 33. This mode of operation is dependent on the transfer of both substrate and product across the semipermeable membrane.

*Enzymes Immobilized with Hollow Fiber Devices*

Using a Dow beaker type device (Bio-Fiber® 50-Dialyzer, mol wt cutoff ca. 5,000), Rony[435] immobilized alkaline phosphatase (mol wt ca. 80,000) and α-chymotrypsin (ca. 25,000) in the

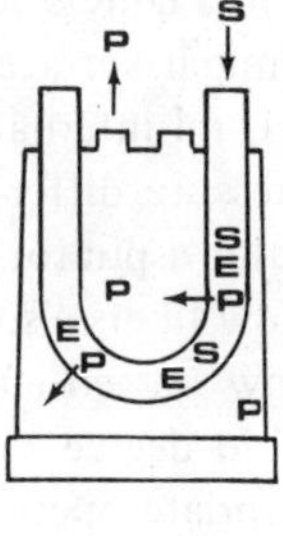


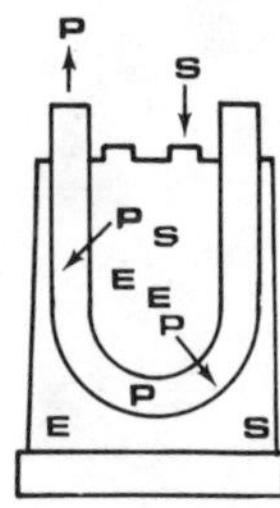


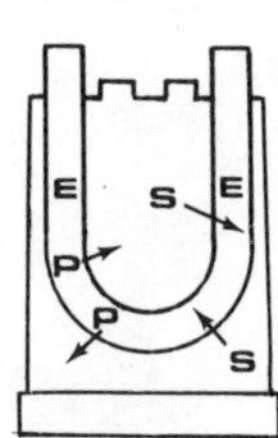


FIGURE 33. Some possible operational modes using the hollow fiber beaker device for continuous enzymatic reactions.

interior of the hollow fibers and examined enzyme leakage, transfer of substrate and product, and activity. He found facile conversion of either *p*-nitrophenylphosphate or BAEE (the substrates for alkaline phosphatase and α-chymotrypsin, respectively) and observed no leakage. Continuous conversion was achieved. Other chemical and physical properties of such immobilized enzymes have not been reported.

*Advantages and Disadvantages*

In addition to the advantages of simplicity, the nonchemical nature of the immobilization, the suitability for the degradation of high molecular weight polymeric substrates, the control of product distribution through membrane cutoff selection, and the simultaneous immobilization of many enzymes, (also mentioned as advantages for ultrafiltration cells), hollow fiber devices have several more advantages. Hollow fibers offer an extremely large surface area to volume ratio, permit continuous operation at lower pressures, and can be used in more varied operational modes. To illustrate these advantages, let us again examine the Dow beaker type device shown in Figure 29. This hollow fiber bundle has a nominal surface area of 1,000 $cm^2$ but only has a 10 ml internal capacity. It normally operates at pressure differences that can be obtained by a water aspirator. The isotropic nature of the Dow hollow fibers also permits solvent and solute to flow in either direction, which allows a considerable degree of freedom in choosing the most appropriate operational mode.

A major disadvantage of the use of hollow fibers for the immobilization of enzymes is, in addition to those mentioned for ultrafiltration cells, the present low availability of different membranes.

**Immobilization with Semipermeable Membranes**

Perhaps the simplest and cheapest device that can be used for immobilizing an enzyme and subsequently employing it as a continuous enzyme reactor is the well-established dialysis tubing most frequently made from cellophane. The enzyme solution is put into the dialysis tubing and the "immobilized" enzyme is dipped into a vessel containing the substrate. A substrate permeable to the membrane can transfer across and be transformed to product. The product, in turn, can transfer across the membrane into the external phase. In essence, this immobilization system is analogous to that of microencapsulation and mode C of hollow fibers. Despite the simplicity of the method, only one example of this approach has been reported. Balcom et al.[100] immobilized catalase with dialysis tubing and attempted to use it for continuous catalysis. Unfortunately, a severe problem with diffusion was encountered and this prevented extended use.

Other examples of membrane immobilization with cellophane[383,390,436,437] and other types of semipermeable materials (polyethylene,[438] Cuprophane,[438] ion-exchangers[439]) have been more successful. The application of membrane-immobilized enzymes for "liquid membrane" electrodes and fuel cells is discussed in Chapter 10.

TABLE 13

**Enzymes Immobilized with Ultrafiltration Cells (Continuous Enzyme Reactors)**

| Enzyme immobilized | Apparatus and membrane employed | Enzyme leakage (if reported) | Ref. |
|---|---|---|---|
| $\alpha$-Amylase (3.2.1.1) | Amicon Model 400 UF cell, PM 10 membrane | Leakage; no leakage after reaching steady state operation | 428, 429 |
| | Amicon Model (402?) UF cell, PM 30 membrane | Not reported | 432 |
| | Amicon Model 52 UF cell, PM 10 membrane | Leakage (extent not reported) | 90 |
| $\alpha$-Amylase (water soluble, carboxymethyl-cellulose-triazinyl derivative) | Amicon Model 52 UF cell, PM 10 membrane | No leakage (leakage prevented) | 90 |
| $\beta$-Amylase (3.2.1.2) | Amicon Model (402?) UF cell, PM 30 membrane | Not reported | 432 |
| Glucoamylase (3.2.1.3) | Amicon Model 400 UF cell, PM 10 membrane | Leakage; no leakage after reaching steady state operation | 427, 429 |
| | Amicon Model (402?) UF cell, PM 30 membrane | Not reported | 432 |
| Glucoamylase (?) ("saccharifying enzyme" | Not specifically reported, membrane equivalent to Diaflo type UM 2 | Not reported | 433 |
| Cellulase (3.2.1.4) | Amicon Model 400 UF cell, HFA 300 membrane | Little or no leakage | 430, 431 |
| | Amicon Model 2000 UF cell, HFA 300 membrane (also PM 30 membrane) | Little or no leakage | 430, 431 |
| Pullulanase (3.2.1.-) | Amicon Model (402?) UF cell, PM 30 membrane | Not reported | 432 |
| $\alpha$-Chymotrypsin (water-soluble, dextran-triazinyl derivative) (3.4.4.5) | Amicon Model 52 UF cell; XM 100 membrane | No leakage (leakage prevented) | 91 |

Chapter 9

# ENZYME REACTORS

There is considerable activity at the moment in the emerging engineering discipline of "enzyme reactors." Good introductory reviews have appeared quite recently on this subject. Barker, Emery, and Novais[83] discussed the characteristics of several types of reactors and emphasized the use of the fluidized-bed reactor. Lilly and Dunnill,[440] on the other hand, presented a more general discussion of biochemical reactors (those using enzymes and microorganisms as the catalysts). Their review is especially instructive. Lilly, Dunnill, and their colleagues have made major contributions in this area and most of what is discussed in this chapter is based on their extensive efforts.

Lilly and Dunnill classified the types of enzyme reactors according to the mode of operation, catalyst retention, and flow characteristics. Another classification of enzyme reactors, not specifically mentioned by them but useful with regard to the present discussion, is based on the physical form of the catalyst. The immobilized enzyme can be either water-soluble (unmodified enzyme or modified derivative) or water-insoluble (an enzyme conjugate).

The operational mode of a reactor may be batch, semibatch, or continuous. Batch operations allow great flexibility, but continuous operations permit diminished labor costs, the facilitation of automatic control, and greater constancy in reaction conditions.

Reactors also may be classified according to the flow characteristics of the system. A reactor in which the flow conditions are uniform is known as a well-mixed or stirred-tank reactor. A plug-flow or tubular reactor is one in which no mixing occurs in the direction of flow; conditions at every point along the reactor are different but constant with time. This latter type is also referred to as a packed-bed or fixed-bed reactor. Although the flow characteristics in reactors can approach the limiting situations of well-mixed and plug-flow, most flow conditions lie somewhere between these extremes.

Catalyst retention in enzyme reactors is an important consideration. A "closed" system, in which catalyst and support are retained, must be maintained for efficient use of enzymes.

Lilly and Dunnill further discussed the factors that determine the choice of reactor type, which in general is dependent on the kinetics of the reaction, the operational requirements, the catalyst replacement or regeneration, reactor costs, and utilization.

This chapter deals mainly with the kinetics of various reactors. Only occasional mention is made of the other factors that can strongly dictate the choice of reactor type. Unfortunately, these factors, especially cost and utilization, are far more difficult to assess objectively and quantitatively than is the kinetics of the reactors. General kinetics of immobilized enzyme-catalyzed reactions is presented in chapters on both particulate- and membrane-based systems and is not discussed here. Good treatments have been presented by Goldman,[261,273] Selegny,[274,275] Sundaram,[441] Kasche,[193] and their respective coworkers.

## Kinetics of Enzyme Reactors

### *Packed-Bed Reactor*

Bar-Eli and Katchalski[55] observed that the extent of hydrolysis of L-arginine methyl ester, poly-L-lysine, protamine, and oxidized insulin by water-insoluble trypsin packed in a column could be regulated by determining the rate of flow of the substrate through the column. They derived an expression relating the effect of enzyme concentration, E, the height of the column, h, the rate of flow, V (in centimeters per unit time), and the substrate concentration, S, with the extent of hydrolysis. Derivation of the mathematical expression was initiated from the Michaelis-Menten relationship previously given in Equation 43 and shown again below in rearranged form (Equation 54).

$$-\frac{dS}{dt} = \frac{kES}{K'_m + S} \qquad (54)$$

The rate constant k is the specific constant

determining the rate of decomposition of the enzyme-substrate complex. The integrated form of Equation 54 is shown in Equation 55.

$$(S_o - S_t) + K'_m \ln \left(\frac{S_o}{S_t}\right) = kEt \quad (55)$$

$S_o$ and $S_t$ are the substrate concentrations at times zero and t. Substituting the time of exposure of the substrate to the enzyme in the column by the ratio of h/V, Bar-Eli and Katchalski obtained Equation 56, in which $S_h$ denotes the concentration of substrate in the effluent.

$$(S_o - S_h) + K'_m \ln \left(\frac{S_o}{S_h}\right) = kE \frac{h}{V} \quad (56)$$

Equation 56 shows that the hydrolysis of substrate is determined by the effective concentration of enzyme, the height of the column, $K'_m$, and the rate of flow of substrate. When $S >> K'_m$, Equation 54 simplifies and upon integration gives Equation 57.

$$\frac{S_o - S_h}{S_o} = \frac{kEh}{S_o V} \quad (57)$$

Equation 57 was verified experimentally for trypsin acting on L-arginine methyl ester. In the presence of a large excess of substrate, the fraction of ester hydrolyzed, $\frac{S_o - S_h}{S_o}$, was found to be inversely proportional to the rate of flow and to the initial concentration of L-arginine methyl ester.

Lilly, Hornby, and Crook[205] likewise derived an equation of the fraction of substrate reacted in the column from the Michaelis-Menten relationship. Substituting the residence time, t, by $\frac{V_l}{Q}$ (the void volume, $V_l$, and flow rate through the column, Q) and the effective enzyme concentration, E, by $\frac{E_T}{V_t}$ ($E_T$ being the total amount of enzyme in moles in the packed bed and $V_t$ the total bed volume) in Equation 55 gives Equation 58.

$$(S_o - S_t) - K'_m \ln \left(\frac{S_t}{S_o}\right) = k \left(\frac{E_T}{Q}\right) \left(\frac{V_\ell}{V_t}\right) \quad (58)$$

Defining P as the fraction of the substrate reacted in the column, $\frac{S_o - S_t}{S_o}$, and $\beta$ as the voidage of the column, $\frac{V_l}{V_t}$, substitution of these respective quantities into Equation 58 gives Equation 59, where $C = kE_T\beta$. C is referred to as the reaction capacity of the column.

$$PS_o - K'_m \ln(1 - P) = \frac{kE_T\beta}{Q} = \frac{C}{Q} \quad (59)$$

Theoretical curves of the relationship between the fraction of substrate reacted in the column and the flow rate (at constant values of C and $K'_m$) defined by Equation 59 are shown in Figure 34. Varying values of initial substrate concentrations are given.

Upon rearrangement, Equation 59 can be transformed to Equation 60.

$$PS_o = K'_m \ln(1 - P) + \frac{C}{Q} \quad (60)$$

This equation shows that if values of P are measured when various initial concentrations of substrate are perfused through the same column at identical flow rates, then $PS_o$ plotted against ln (1 − P) will give a straight line provided $K'_m$ and C are constant at this constant flow rate. The slope of the line is equal to $K'_m$ and the intercept on the $PS_o$ axis is equal to $\frac{C}{Q}$. The apparent Michaelis constant and the reaction capacity can be determined for any flow rate of substrate through the column by this plotting technique.

Lilly, Hornby, and Crook[205] investigated also the validity of Equation 59 (or its rearranged

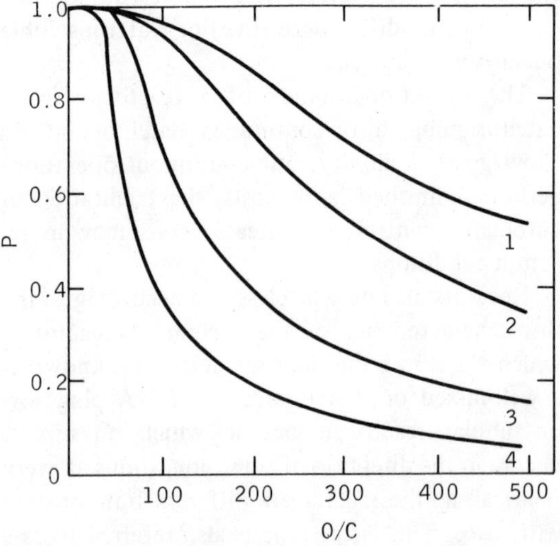


FIGURE 34. Theoretical relationships between degree of reaction, P, column reaction capacity, C, and flow rate, Q, for $K'_m$ 2.5 mM. Curves for $S_o = 10^{-1}$ $K'_m$ (1), $S_o = K'_m$ (2), $S_o = 4$ $K'_m$ (3), and $S_o = 10$ $K'_m$ (4) are shown. (Redrawn from Lilly, M. D., Hornby, W. E., and Crook, E. M., *Biochem. J.*, 100, 718, 1966. With permission.[205])

counterpart, 60) with ficin covalently bonded to several CM-celluloses of different ion-exchange capacities. Some of their results are shown in Figures 35 to 37.

Figure 35 gives the experimental curve of the fraction of substrate reacted in the column, P, plotted against the flow rate, Q, and it can be seen that the general shape of the curve is that expected from Equation 59. However, the expected decrease in the slope at low flow rates (see Figure 34) was not observed with this or any other ficin-CM-cellulose derivative acting on BAEE. The expected behavior (decreased slope with decreased flow rates) was noted, however, in the hydrolysis of casein (figure not shown), which has a higher column reaction capacity than BAEE.

Figure 36 gives the relationship between $PS_o$ and log (1 – P) obtained for the hydrolysis of BAEE by the immobilized ficin and shows that the flow rate can have a pronounced effect on $K'_m$. If Equation 59 were to be applicable under all conditions, the parameters $K'_m$ and C would have to be independent of flow rate and of initial concentration. No systematic effect was, however, discerned by Lilly et al. In all ficin derivatives

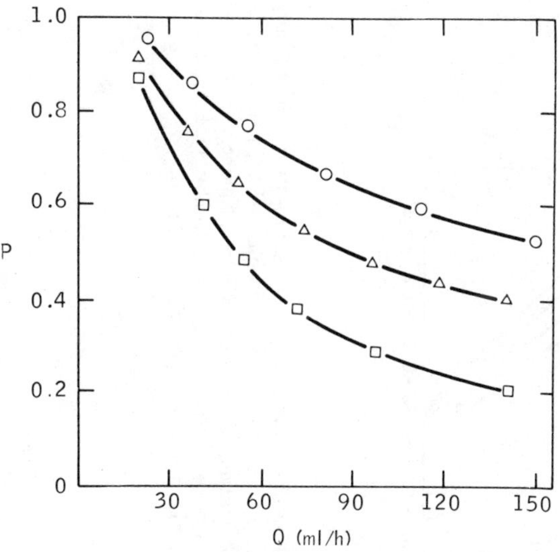


FIGURE 35. Relation between degree of hydrolysis, P, of BAEE and flow rate, Q, for a column of CM-cellulose-ficin with initial substrate concentrations 0.5 m*M* (○), 5 m*M* (△), and 15 m*M* (□) in phosphate-NaCl (ionic strength 0.4). Exchange capacity of 0.90 mequiv/g for CM-cellulose. (Redrawn from Lilly, M. D., Hornby, W. E., and Crook, E. M., *Biochem. J.*, 100, 718, 1966. With permission.[205])

examined, $K'_m$ increased at low flow rates, the increase being dependent on the ionic strength of the mixture and the degree of substitution (ion-exchange capacity) of the cellulosic support (see Figure 37). At high flow rates, the apparent Michaelis constant in general approached, but never reached, the value observed under comparable conditions in stirred suspensions. An exception to the above behavior was noted for ficin attached to CM-cellulose (having an ion-exchange capacity of 0.90 mequiv/g) at high ionic strengths. As the flow rate increased, $K'_m$ decreased and tended toward the value of $K'_m$ obtained in stirred suspensions. This dependence of $K'_m$ on flow rate (at low flow rates) was explained in the following way. Each ficin-cellulose particle is surrounded by a diffusion layer, and the transfer of substrate through this layer is inversely related to the thickness of the layer. In addition, the effective thickness of the diffusion layer is inversely related to the flow of solution past the immobilized enzyme particle. This means that the rate of diffusion of substrate to the enzyme is dependent on the flow rate. Consequently, at lower flow rates, there is a significant difference between the concentration of substrate in the bulk of the solution and in the immediate vicinity of the immobilized enzyme. This difference is manifested in an elevation of the apparent $K'_m$.

Values of C and k for ficin-CM-cellulose derivatives were also reported. The reaction capacity was found to be nearly constant with flow rate. Values of k for the ficin-cellulose derivatives in a column were less than the values reported for the same preparations in stirred suspensions. At low ionic strength, the value of k for ficin bonded to CM-cellulose (the material with 0.90 mequiv/g ion-exchange capacity) did, however, approach the value observed in stirred mixtures. A plausible explanation for this apparent loss in activity, suggested by Lilly et al., could be the inaccessibility of some of the ficin molecules to substrate molecules because of the much closer proximity of the cellulose fibers inherent in the packing of the column. This, in turn, could give rise to stagnant interfibrous regions that form a barrier between substrate and those enzyme molecules located within this zone. The result would be a decrease in the effective concentration of the enzyme.

Recently, Tosa et al.[324,352] reported another variation of the Michaelis-Menten-derived analysis for the kinetic behavior of immobilized enzymes

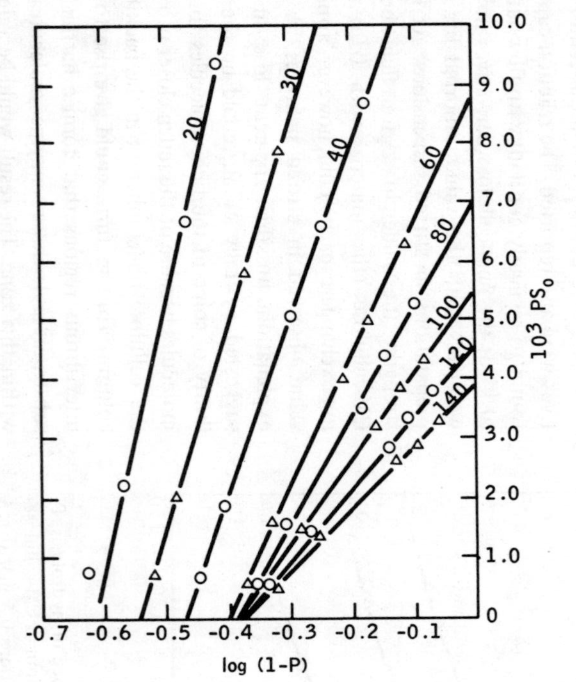


FIGURE 36. Relationship between $PS_o$ and log (1 – P) for the hydrolysis of BAEE by CM-cellulose-ficin in a column (phosphate-NaCl, ionic strength 0.4). Exchange capacity of 0.90 mequiv/g for CM-cellulose. Numbers on the curves give flow rate, Q, ml/hr. (Redrawn from Lilly, M. D., Hornby, W. E., and Crook, E. M., *Biochem. J.,* 100, 718, 1966. With permission.[205])

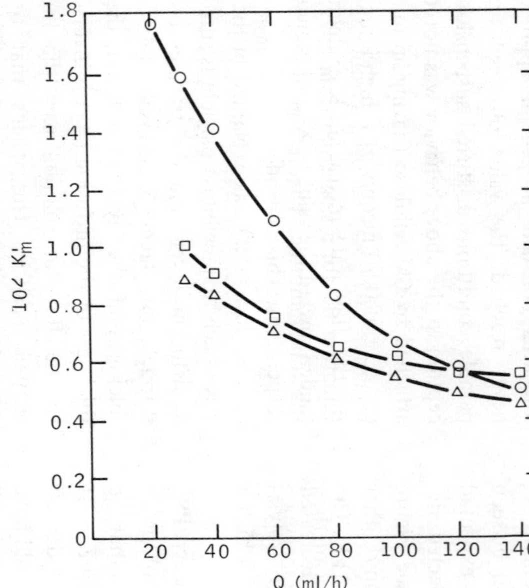


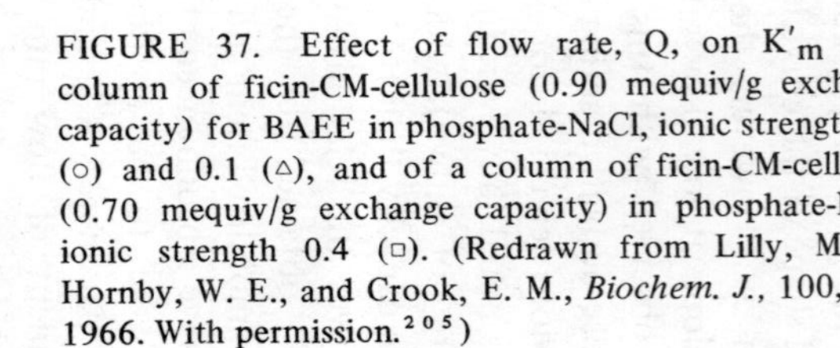

FIGURE 37. Effect of flow rate, Q, on $K'_m$ of a column of ficin-CM-cellulose (0.90 mequiv/g exchange capacity) for BAEE in phosphate-NaCl, ionic strength 0.4 (○) and 0.1 (△), and of a column of ficin-CM-cellulose (0.70 mequiv/g exchange capacity) in phosphate-NaCl, ionic strength 0.4 (□). (Redrawn from Lilly, M. D., Hornby, W. E., and Crook, E. M., *Biochem. J.*, 100, 718, 1966. With permission.[205])

in columns. Their analysis was based on space velocity, $V_s$, which is defined as the volume of liquid passing through a given volume of water-insoluble enzyme in one hour divided by the latter volume. Because $\frac{h}{V}$ is equal to $\frac{1}{V_s}$, Equation 56 can be transformed, by substitution, into Equation 61. Other variations of this equation, similar to those developed above by Lilly et al.,[205] were also given.

$$(S_o - S_h) + K'_m \ln\left(\frac{S_o}{S_h}\right) = \frac{kE}{V_s} \tag{61}$$

Tosa et al.[324,352] tested the validity of their space velocity equations with aminoacylase adsorbed on DEAE-Sephadex and DEAE-cellulose. In the presence of a relatively large excess of the substrate *N*-acetyl-DL-methionine, the fraction hydrolyzed was found to be inversely proportional to the space velocity and independent of the height of the enzyme column. Plots of log (1 - P) vs. $PS_o$ were likewise linear and dependent on flow rates. As the flow rate was increased, the value of the apparent Michaelis constant decreased. The effect of flow rate on k was reported to be negligible, but a definite downward trend was observed with increased flow rates.

The validity of Equation 60 was also examined for packed-bed reactors of amyloglucosidase[65] covalently bound to DEAE-cellulose and for saccharase[298] adsorbed on DEAE-cellulose.

A special case of packed-bed reactors is that made of porous sheets to which enzymes have been attached. The kinetic properties of these packed-bed reactors have been examined by Lilly and several of his co-workers.[40,62,64] The kinetic behavior of systems such as β-galactosidase[64] covalently bonded to cellulose or DEAE-cellulose sheets and pyruvate kinase[40] bound to cellulose seems to obey the relationship given in Equation 60. With β-galactosidase, linear relationships between $PS_o$ and ln (1 - P) were observed, but the values of $K'_m$ and C changed with flow rate through the porous sheets depending on the substrate. With lactose as substrate, C remained almost constant but $K'_m$ decreased with flow rate (except at 5°). With *o*-nitrophenyl galactoside as the substrate, C increased with flow rate but $K'_m$ varied far less than it did with lactose.

The effect of substrate inhibition has been examined theoretically and experimentally. Wilson, Kay, and Lilly[63] derived an expression relating the degree of substrate conversion of an immobilized enzyme exhibiting substrate inhibition with the parameters $K'_m$, C, Q, and the substrate inhibition constant $K_s$. Equation 62 is their theoretically derived expression for a column of insoluble substrate-inhibited enzyme with $K'_s$ being the apparent substrate inhibition constant.

$$S_o - S_t = K'_m \ln\left(\frac{S_t}{S_o}\right) + \frac{C}{Q} - \left(\frac{1}{2K'_s}\right)(S_o^2 - S_t^2) \tag{62}$$

Equation 62 was derived from the kinetic Equation 63 for a single substrate enzyme that is substrate-inhibited. Theoretical curves for various values of $K'_s$, $K'_m$, and $\frac{C}{Q}$ were computed and given in graphical form (not shown).

$$v = \frac{kE}{1 + \frac{K_m}{S_o} + \frac{S_o}{K_s}} \tag{63}$$

Wilson et al.[63] examined experimentally the validity of Equation 62 for lactate dehydrogenase covalently bonded to DEAE-cellulose sheets and observed that this enzymic system, at least for low values of $S_o$, did not obey the equation. A considerable difference existed between the theoretical and experimental curves of $\ln(\frac{S_t}{S_o})$ against $(S_o - S_t)$. The variation of $K'_m$ and C with flow rate was also examined and was noted to be different from that previously observed for ficin attached to CM-celluloses. An increase in the flow rate for lactate dehydrogenase-DEAE-cellulose sheets caused an increase in the value of the reaction capacity but not in the apparent Michaelis constant, which remained essentially constant over the flow rates employed. The occurrence of substrate inhibition in a packed-bed reactor was suggested for penicillin amidase[66] covalently attached to DEAE-cellulose. In this system, however, the possibility of product inhibition was present and this could have been responsible (at least in part) for the observed deviations from pure Michaelis-Menten kinetics.

*Continuous-Feed Stirred-Tank Reactors*

A kinetic analysis of continuous-feed stirred-tank (CFST) reactors employing water-insoluble enzyme derivatives was given by Lilly and Sharp[189] in 1968. Equation 64, relating the fraction of substrate converted to product in such a reactor with $K'_m$, C, and Q, was derived in a fashion analogous to that for Equation 60.

$$PS_o + \left(\frac{P}{1 - P}\right) K'_m = \frac{kE_T\beta}{Q} = \frac{C}{Q} \tag{64}$$

Perfect mixing throughout the vessel is assumed. Derivation of Equation 64 followed from a consideration of the substrate balance between input, output, and substrate conversion (Equation 65).

$$QS_o = QS_i + \frac{kES_iU_L}{K'_m + S_i} \tag{65}$$

$S_i$ is the substrate concentration and $U_L$ is the liquid volume in the stirred tank. The enzyme concentration in this solid-liquid system is defined by $E = \frac{E_T}{U_T}$ where $U_T$ is the volume of the liquid and enzyme derivative in the stirred tank. Rearranging Equation 65 and introducing the above relationship of E gives Equation 66.

$$Q(S_o - S_i) = \frac{k\left(\frac{E_T}{U_T}\right) U_L}{\frac{K'_m}{S_i} + 1} \tag{66}$$

By defining $P = \left(1 - \frac{S_i}{S_o}\right)$ and the voidage of the reactor as $\frac{U_L}{U_T}$, Equation 66 may be written as Equation 67. Upon rearrangement, Equation 67 becomes Equation 64.

$$PS_oQ = \frac{kE_T\beta}{\frac{K'_m}{S_o(1 - P)} + 1} \tag{67}$$

Theoretically derived curves for the variation of the reaction capacity with input substrate concentration at various values of conversion of substrate and apparent Michaelis constant at constant flow rates were given (not shown).

The validity of Equation 64 for an immobilized enzyme-catalyzed system in a continuous-feed stirred-tank reactor was examined recently by O'Neill, Dunnill, and Lilly[151] with amyloglucosidase covalently bonded to DEAE-cellulose. These investigators followed the conversion of maltose to glucose at constant flow rate, temperature, and operating tank volume and found that the reaction rate in the CFST reactor followed Michaelis-Menten kinetics and that Equation 64 was a suitable model for this reactor system. The value of $K'_m$ obtained in the CFST reactor study (1.1 m*M*) was close to the value previously obtained from initial-rate batch data ($K'_m$ = 1.4 m*M*).

The effect of substrate inhibition on the kinetic behavior of insolubilized enzymes in continuous-feed stirred-tank reactors has also been considered. O'Neill, Lilly, and Rowe[442] recently derived Equation 68 from the equation for a single substrate enzyme that is substrate-inhibited (Equation 63).

$$PS_o + K'_m \left(\frac{P}{1 - P}\right) + \frac{S_o^2}{K'_s} P(1 - P) = \frac{kE_T}{Q} \tag{68}$$

The theoretical occurrence and significance of multiple steady states for substrate-inhibited enzymes in CFST reactors was also discussed by these investigators.[442]

*Comparison of the Packed-Bed and Continuous-Feed Stirred-Tank Reactor*

Theoretical and experimental comparisons of packed-bed and continuous-feed stirred-tank reactors have been made by the British workers.[151,189,442,443] Lilly and Sharp[189] compared these two types of reactors theoretically and reported that under most conditions the packed-bed reactor is more efficient on the basis of product output per unit of enzyme than the continuous-feed stirred-tank reactor. The CFST reactor is, however, more efficient at low values of substrate conversion and at high substrate concentrations. Likewise, the CFST reactor becomes more favorable than the packed-bed reactor when substrate inhibition is present.[151,440] On the other hand, when product inhibition occurs, the CFST reactor is even less efficient than the packed-bed reactor.[440] Diffusion limitation may be significant in either type of reactor.[443]

Other theoretical comparisons of these two limiting cases of reactor types have also appeared. Rony[434] calculated concentration profiles, apparent Michaelis constants, effectiveness factors, and reactant conversions for both a fixed-bed and CFST reactor. A theory of diffusion control was derived for coordinate geometries that corresponded to the "liquid-filled hollow plane membrane," the hollow fiber, and the microcapsule. The effects of dispersion and mass transfer resistance on the degree of conversion in immobilized enzyme reactors were considered mathematically by Kobayashi and Moo-Young.[444]

A good example of the examination of the behavior of an immobilized enzyme in these two types of reactors was given by O'Neill, Dunnill, and Lilly.[151] The kinetic behavior of amyloglucosidase covalently bonded to DEAE-cellulose was examined in a packed-bed reactor and a continuous-feed stirred-tank reactor for the conversion of maltose to glucose. Some of the results of the CFST reactor have already been described in connection with testing the validity of Equation 64, and it will be recalled that water-insoluble amyloglucosidase did, indeed, show kinetic behavior completely consistent with this equation. The same water-insoluble derivative was packed into a bed and its behavior likewise examined.

Substrate conversion as a function of flow rate in the two reactors under identical operating conditions is shown in Figure 38. The predicted packed-bed reactor relationship of P vs. Q (using Equation 60 and the values of $K'_m$ and $kE_T$ obtained in the CFST reactor) is also shown. At all flow rates used, the CFST reactor surprisingly, and contrary to theoretical considerations, shows a substrate conversion equal to or greater than the packed-bed reactor. Analysis of these reactor systems by O'Neill et al. revealed that the unexpected low efficiency of the bed reactor was the result of poor mass transfer due to the inability of the reactor liquid to completely permeate the

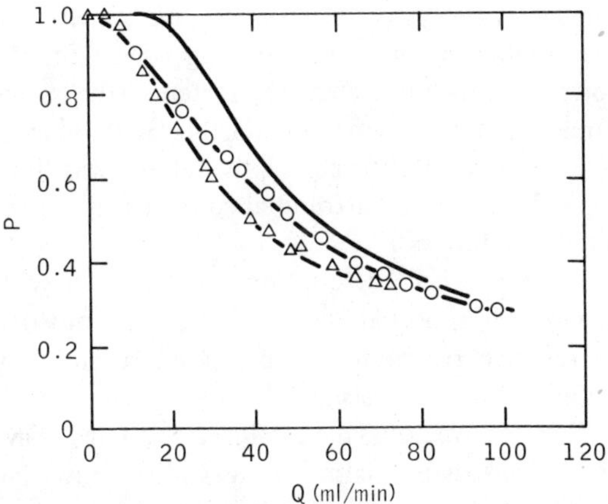


FIGURE 38. Relation between substrate conversion, P, and flow rate, Q, for a CFST (○) and packed-bed (△) reactor with identical operating parameters. $S_o$ = 5 m*M* maltose; sodium acetate buffer pH 4.3, 25°; same amount of enzyme. Upper solid line represents predicted packed-bed reactor conversion using equation 60 with values of $K'_m$ and $kE_T$ taken from CFST experiment. (Redrawn from O'Neill, S. P., Dunnill, P., and Lilly, M. D., *Biotechnol. Bioeng.*, 13, 337, 1971. With permission.[151])

cellulose support structure of the packed bed. It was not, however, attributed to film diffusional resistances at all catalytic sites in the bed. Equation 60 was not valid for the immobilized amyloglucosidase packed bed due to this lack of solid-fluid contact in the catalyst bed – a lack which decreased the efficiency of the bed reactor to such an extent as to offset the inherent kinetic advantages of the bed reactor compared with the CFST reactor.

**Types of Reactors Used**

Various types of reactors have been used with soluble and insoluble enzymes. Most frequently employed is the plug-flow reactor using water-insoluble enzyme conjugates that are either in particulate or sheet form. These have been used for various enzymes and water-insoluble supports and for operations of widely different scales. Some rather well-described column operations of immobilized enzymes can be found in the literature.[55,65,132,145,151,205,298,304,324,352] Hornby and Filippusson[101] also described the preparation and characterization of enzymes covalently bonded to the inside surface of nylon tubes and the use of such hollow fibers as enzyme reactors. Hindered-settling or expanded-bed reactors have likewise been mentioned as enzyme reactors.[151,440] A modular reactor has been reported for collagen-enzyme conjugates.[293,342]

Continuous stirred-tank reactors have been employed often as enzyme reactors either in batch or continuous operational modes. Both particle-bound[146,151,343] and water-soluble[91,428-431] enzymes (either unmodified or derivatized) have been used. The use of ultrafiltration cells as reaction vessels should increase the prevalence of this type of reactor in the future.

The use of fluidized-bed reactors has been reported with water-insoluble enzyme derivatives.[83]

An interesting continuous enzyme reactor was described by Reese and Mandels.[445] Their reactor consisted of a two-phase column, the operation of which was based on partition chromatography. In their method, the enzyme (dissolved in an aqueous phase) was retained as the stationary phase on a column of a hydrophilic solid such as cellulose. The substrate, moving through the column in the solvent phase, diffused into the stationary aqueous phase wherein the reaction with the enzyme occurred. Products of the reaction diffused into the mobile phase and eventually passed down the column. The method was successfully used to achieve continuous enzymic conversion using β-glucosidase and invertase. An important feature of this reactor is that the enzyme is not retained (immobilized) by physical adsorption. The role of the solid phase in the column is simply mechanical; it holds the aqueous phase in position and creates a larger interfacial surface area.

**Consecutive Enzyme-Catalyzed Systems**

An important consideration for biological systems and enzyme reactors is the nature of consecutive reactions catalyzed by several enzymes. These are systems in which the product of one enzyme-catalyzed reaction is the substrate for another enzyme in a sequence (Equation 69).

$$A \xrightarrow{E_1} B \xrightarrow{E_2} C \xrightarrow{E_3} D \longrightarrow \quad \longrightarrow \qquad (69)$$

An understanding of the behavior of these reactions is highly desirable in order to gain an insight into the exact sequence of events during metabolic processes and to create more efficient and complex enzyme reactors.

Several synthetically prepared consecutive enzyme-catalyzed reaction systems have been described. These have been constructed either by simultaneous or sequential immobilization. Simultaneous immobilization is the localization of two or more enzymes at the same time and in the same micro-vicinity. It may be achieved by such means as covalent bonding of the enzymes to the same functionalized polymer or by entrapment of the catalysts within the same microspace of a crosslinked polymer, microcapsule, or membrane-dependent device. Sequential immobilization is the mutually independent localization of two or more enzymes, achieved by immobilizing the enzymes individually (one at a time) and then arranging them into an ordered array.

Simultaneous immobilization of two or more enzymes has been achieved by several different methods. Hexokinase[107] and glucose-6-phosphate dehydrogenase were immobilized simultaneously by (1) polyacrylamide gel entrapment and by (2) covalent bond formation to Sepharose and to the copolymer of acrylamide and acrylic acid. β-Galactosidase,[109] hexokinase, and glucose-6-phosphate dehydrogenase were similarly covalently bonded to CNBr-activated Sephadex G-50 C. Glucose oxidase[369] and peroxidase were immobilized by lattice-entrapment with Silastic resin. Crude multi-enzyme systems[406] capable of converting glucose to either ethanol or lactate were immobilized by entrapment within silicone-based microcapsules.

Kinetic characterization of the hexokinase-glucose-6-phosphate dehydrogenase system and the ternary system prepared with β-galactosidase revealed some noteworthy information. Mosbach and Mattiasson[107] observed an increased rate of formation of NADPH for all two-enzyme systems of hexokinase and glucose-6-phosphate dehydrogenase (Equation 70). Taking the number of moles of NADPH produced/min/ml on the freely soluble system as reference (100%), an increase in the formation of 40 to 100 and 140% was observed for the Sepharose and copolymeric acrylamide-acrylic acid systems, respectively. The polyacrylamide lattice-entrapped system similarly showed a 110% increased NADPH formation. The enhanced formation of reduced coenzyme was attributed to higher local concentrations of glucose-6-phosphate in the immediate vicinity of the immobilized glucose-6-phosphate dehydrogenase compared with that of the freely soluble systems. The ternary[109] system of β-galactosidase, hexokinase, and glucose-6-phosphate dehydrogenase (Equation 71) also exhibited enhanced NADPH formation relative to the freely soluble system but in addition showed a cumulative effect. The time required for the immobilized three-enzyme system to approach the activities of the soluble systems was far longer than for the corresponding two-enzyme system (the same three-enzyme system but starting with glucose transformation). The cumulative effect was time-dependent and was attributed to the nature of the sequential process.

A theoretical description of a two-enzyme membrane-supported system that carries out two consecutive reactions was presented recently by Goldman and Katchalski.[273]

Consecutive enzyme-catalyzed systems have been constructed also from individually immobilized enzymes. For example, Brown et al.[363] immobilized separately hexokinase, phosphoglucoisomerase, phosphofructokinase, and aldolase by lattice-entrapment within polyacrylamide and arranged the enzyme conjugates in a column in the same order as found in nature (Equation 72). A substrate solution containing 1 mmole glucose, 0.3 mmole ATP, and 1 mmole $MgCl_2$ was placed on the column and eluted fractions were assayed for

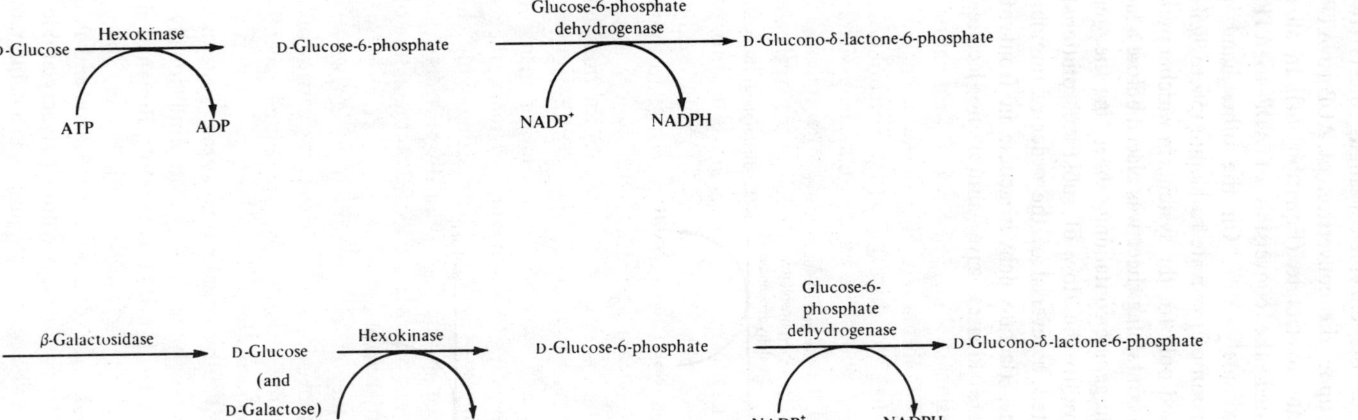


(70)

(71)

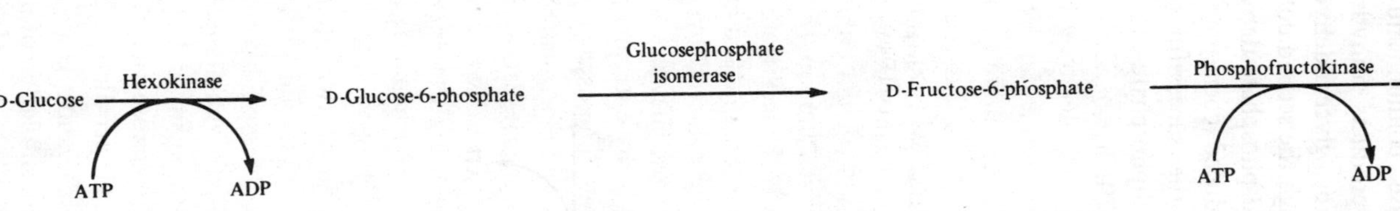


$$\text{D-Fructose-1,6-diphosphate} \xrightarrow{\text{Aldolase}} \text{D-Glyceraldehyde-3-phosphate} + \text{Dihydroxyacetone phosphate}$$

(72)

total unreacted glucose, inorganic phosphate, and glyceraldehyde-3-phosphate. Analysis showed that 116 μmoles of glucose were consumed and that approximately 12 μmoles of glyceraldehyde-3-phosphate were produced.

A two-enzyme system consisting of aldolase[116] and glyceraldehyde-3-phosphate dehydrogenase, immobilized by covalent bonding to AE-cellulose with glutaraldehyde, was likewise used in a packed-bed reactor (Equation 73). A solution of fructose-1,6-diphosphate and $NAD^+$ was first passed through the aldolase column and the effluent (containing the products dihydroxyacetone phosphate and glyceraldehyde-3-phosphate) was passed through the second column packed with glyceraldehyde-3-phosphate dehydrogenase. The presence of reduced coenzyme was taken as an indication of the successful transformation of fructose-1,6-diphosphate to the products shown in Equation 73.

A two-enzyme packed-bed reactor consisting of pyruvate kinase and lactate dehydrogenase covalently bonded to separate cellulose sheets was described by Wilson, Kay, and Lilly.[40] These workers investigated the effect of varying the order of the enzyme-containing sheets (two of each) upon the conversion of ADP to ATP and pyruvate to lactate (Equation 74). In all cases examined, the conversion of ADP to ATP was approximately 30%. On the other hand, good conversion of pyruvate to lactate (59 to 97%) was observed only for the systems in which a pyruvate kinase-containing sheet was placed before a lactate dehydrogenase-containing sheet in the reactor (relative to the flow of substrate solutions). As expected, a reversal of the order of immobilized enzymes (lactate dehydrogenase in front of the pyruvate kinase) gave almost no lactate (3% conversion).

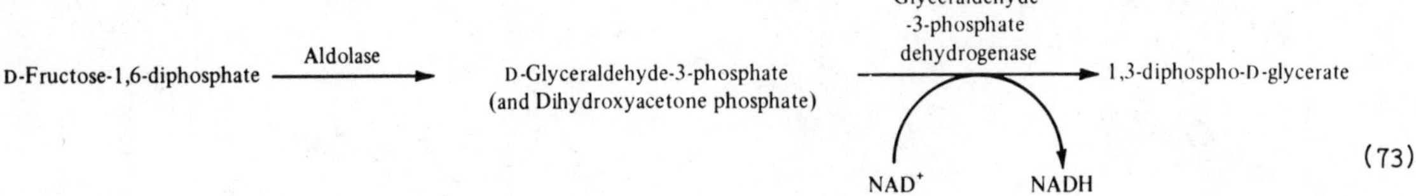


(73)

Phosphoenolpyruvate —Pyruvate kinase (ADP → ATP)→ Pyruvate —Lactate dehydrogenase (NADH → $NAD^+$)→ Lactate (74)

# Chapter 10

# APPLICATIONS

Applications of immobilized enzymes can be classified into two general categories, depending on the two major characteristics of enzymes – their catalytic properties and their structural properties. Table 14 gives the classification system adopted and lists the general areas of application. The catalytic property-dependent areas are largely arbitrarily selected, and often a given application could equally well be placed in several areas. Applications discussed in these sections fall predominantly into the chemical, pharmaceutical, medical, and food industries.

TABLE 14

**General Application Areas of Immobilized Enzymes**

| Areas dependent on catalytic properties | Areas dependent on structural properties |
|---|---|
| • Chemical processes | • Separations |
| • Analysis | |
| • Mechanistic studies (model systems) | |
| • Metabolic disorders | |
| • Fuel cells | |
| • Facilitated transport | |

## Applications Dependent on Catalytic Properties of Enzymes

### *Chemical Processes*

It is convenient to discuss the applications of this section in terms of the enzyme classification system. Thus, enzymes carry out oxidation-reduction, group transfer, hydrolytic, addition, isomerization, and condensation reactions. An additional group of reactions that could be included is multi-step enzymic reactions (see Chapter 9). The emphasis of this section is on the conversion of reactant to product whether it is for obtaining a desired product or for eliminating a reactant due to its objectionable character.

#### ***Oxidation-Reduction Reactions***

Several examples of chemical conversions using immobilized enzymes of this group are known. A steroid dehydrogenase[379] from *Corynebacterium simplex* that catalyzes the $\Delta^1$-dehydrogenation of cortisol to prednisolone, immobilized by lattice-entrapment, gave good conversions of substrate to product in a continuous column operation. L-Amino acid oxidase,[121] isolated from *Crotalus adamanteus* and immobilized on glass, was used for both the resolution of racemic mixtures of amino acids and for the preparation of amino acids. Weetall and Baum[121] converted a racemic mixture of DL-phenylalanine with the immobilized enzyme, in both a recirculating and continuous addition column operation, to a mixture showing optical activity but still containing considerable amounts of unreacted L-phenylalanine. The immobilized L-amino acid oxidase also was used to produce tyrosine by the transamination of DL-phenylalanine and *p*-hydroxyphenylpyruvic acid. A closed-column feed system gave a 2% conversion to tyrosine above a 2 $\mu M$ *p*-hydroxyphenylpyruvic acid concentration (Equation 75). Although the same percent conversion was reported using the freely soluble enzyme, it was estimated that the apparent conversion rate for the bound enzyme in the column was about 10 times greater than for the free enzyme. The conversion of L-tyrosine to 3,4-dihydroxyphenylalanine (L-DOPA) using immobilized mushroom tyrosinase was investigated by Wykes et al.[122] as a possible method for large-scale synthesis of this anti-Parkinson's disease drug. Columns of the enzyme conjugate were used to convert buffered solutions of tyrosine to L-DOPA in the presence of ascorbate, which was added to reduce the formed quinone and to prevent further oxidation of L-DOPA. Unfortunately, both the free and immobilized enzymes exhibited significant and irreversible inhibition that ruled out the use of mushroom tyrosinase for L-DOPA production in an industrial process. Tyrosinases from other sources, especially mammalian and not subject to such inhibition, would be more suitable.

#### ***Group Transfer Reactions***

A few immobilized enzymes of this group have

$$\text{Phenylalanine} + p\text{-Hydroxyphenylpyruvic acid} \underset{\text{(Anaerobic conditions)}}{\overset{\text{L-Amino acid oxidase}}{\rightleftharpoons}} \text{Phenylpyruvic acid} + \text{Tyrosine} \qquad (75)$$

been suggested or used in conversion applications. Dextransucrase[318] adsorbed on DEAE-Sephadex A-50 was found to catalyze the polymerization of sucrose into a fructose-rich polymer that had the ability to interact with concanavalin A. The polymer formed by the immobilized enzyme exhibited a higher glycogen value and fructose content than the polymer synthesized by the water-soluble enzyme. An enzymatic synthesis of $NAD^+$ using NAD pyrophosphorylase adsorbed on hydroxylapatite was reported by Traub et al.[322] A continuous conversion of NMN and ATP to $NAD^+$ and inorganic pyrophosphate, respectively, was achieved in a column operation. The maximum percentage of substrate conversion to $NAD^+$ was reported to be approximately 80%. Polynucleotide phosphorylase[129,295] has also been immobilized and used for preparative purposes. Hoffman et al.[129] immobilized the enzyme on cellulose and employed the covalently bound conjugate intermittently in a total of 40 polymerization cycles. As with the free enzyme, short time cycles afforded polymers of high molecular weight while long incubations (overnight) effected a redistribution of chain lengths and reduced the average sedimentation coefficient. Poly A, U, I, and C were prepared by the immobilized enzyme from the corresponding diphosphates. The biological activities and physical characteristics of poly I and poly C prepared from these homopolymers were indistinguishable from the polymers prepared with the soluble enzyme. Thang et al.[295] had previously demonstrated the polymerization of the ADP into a high molecular weight poly A. The enzyme was immobilized by adsorption on a nitrocellulose Millipore® filter; repeated use of the enzyme conjugate was possible.

### *Hydrolyses*

The hydrolases constitute the largest single group of enzymes that have been immobilized and used for the various applications listed in Table 14. They have been frequently used for chemical conversions because they are easy to obtain in pure form, are stable, and require no coenzymes. It is beyond the scope of this book to give all examples of their use; only a representative number of immobilized enzymes and their applications are discussed. Data pertaining to the application of several of these immobilized hydrolases are discussed in Chapters 8 and 9. References cited here are limited to those that describe the operation (usually a continuous conversion) in some detail.

*α*-Amylase hydrolyzes the *α*-1,4-glucan link in polysaccharides to produce oligosaccharides together with variable amounts of glucose, maltose, and maltotriose, depending on the source of the enzyme. Immobilized *α*-amylase has been utilized for the continuous hydrolysis of starch in both soluble[90,428,429] and insoluble forms.[146] Butterworth et al.[428,429] used a continuously stirred ultrafiltration cell for the conversion of the polysaccharide and observed that the molecular weight distribution of the products was dependent on the nominal molecular weight cutoff of the membrane, absolute ultrafiltration pressure, enzyme-to-substrate ratio, temperature, and residence time of the starch in the reactor. A dextrose equivalent of 34 was reported for the filtrate during steady state operations (dextrose equivalent of maltose is 50). Wykes et al.[90] similarly used an ultrafiltration cell for the continuous hydrolysis of starch. However, a water-soluble chemically modified derivative of *α*-amylase (covalently bonded to CM-cellulose and exhibiting superior thermal stability) was employed. After reaching a steady state, the derivatized α-amylase produced half the amount of hydrolyzed starch compared with the maximum amount produced by the freely soluble enzyme (same protein content in each reactor). The unmodified enzyme lost 82% of its activity after 70 hr compared with the α-amylase-cellulose conjugate, which lost 32% of its activity in the same amount of time. The relative activity of the α-amylase derivative used in the continuous conversion was 10%. Water-insoluble derivatives of α-amylase were employed by Ledingham and Hornby[146] in a continuous stirred reactor for the hydrolysis of amylose. Surprisingly, each derivative of the enzyme (attached to CM-cellulose, *p*-aminobenzylcellulose, and polystyrene) exhibited a higher degree of multiple attack than the unmodified water-soluble enzyme. The term "multiplicity of attack" is used to describe the number of catalytic events (i.e., the number of α-1,4-glucan links hydrolyzed) that occur on a single amylose chain during the lifetime of the enzyme-substrate complex. A marked increase in the production of glucose was also observed for the polystyrene derivative. The increase in the multiplicity of attack was attributed to a steric effect arising from the particulate nature of the supports.

β-Amylase[432] has been employed for the continuous hydrolysis of starch in an ultrafiltration cell. β-Amylase also hydrolyzes α-1,4-glucan links in polysaccharides, but degradation occurs by removal of successive maltose units from the nonreducing ends of the polysaccharide chains. The hydrolytic products of potato starch consisted mainly of the expected product maltose.

γ-Amylase, more commonly known as glucoamylase or amyloglucosidase, has been employed frequently for the hydrolysis of starch. Like α- and β-amylase, glucoamylase hydrolyzes α-1,4-glucan links in polysaccharides, but hydrolysis occurs by the removal of successive glucose units from the nonreducing ends of the polysaccharide chain. The native enzyme has been used for continuous conversions in ultrafiltration cells[427,429,432] and derivatized water-insoluble conjugates have been used in packed-bed[65,151,302] continuous-feed stirred-tank[151,343] and fluidized-bed[83] reactors.

Another enzyme, closely related to the amylases, has been used for continuous conversions both by conventional and by somewhat novel means. Cellulase, the enzyme that hydrolyzes cellulose by breaking β-1,4-glucan links, was employed in stirred-tank,[304,431] ultrafiltration cell,[430,431] and column[304] reactors. The novel systems alluded to consisted of stirred-tank or packed-bed reactors that utilized cellulose, the substrate, as the immobilization support. Cellulose strongly adsorbs cellulase at conditions that are optimum for enzyme activity (pH 4 to 5 and 25 to 50°). The adsorbed enzyme was used to digest cellulose with no requirement of enzyme replenishment even though the liquid phase containing the glucose was continuously removed. As cellulose was digested, the released enzyme was simply readsorbed on excess or newly added cellulose. Syrups containing 5 to 14% glucose were produced continuously from stirred-tank reactors containing 10 to 20% cellulose or from cellulose columns.

Other immobilized hydrolases have been used or proposed for various chemical processing applications. β-Fructofuranosidase[300] was used for the hydrolysis of sucrose, carboxypeptidase[36] for hydrolysis and subsequent resolution of amino acids, pepsin[162] for the continuous coagulation of skim milk, papain[280] for chillproofing beer, pronase[215] for reducing the antigenicity of 6-aminopenicillanic acid (6-APA) preparations, and ATP[301] deaminase for IMP formation from AMP. Applications for immobilized naringinase[156] and penicillin amidase[66] have been proposed. Naringinases are fungal enzymes that hydrolyze naringin, the bitter 7β-neohesperidoside of naringinin that is primarily responsible for the bitterness of grapefruit. Hence, the debittering of clarified juices could be accomplished either in batch or continuous operations using the immobilized enzyme. Immobilized penicillin amidase[66] was investigated in the laboratory and suggested for possible large-scale use in hydrolyzing penicillins to 6-APA.

The most dramatic example to date of the successful use of an immobilized enzyme is aminoacylase hydrolysis of *N*-acyl-L-amino acids reported by Tosa, Chibata, Mori, and their colleagues.[225,291,292,311,316,319,324,325,352] In a series of papers, these investigators described the preparation of various water-insoluble aminoacylase conjugates by either adsorption[291,292,311,316,319,324,325,352] or covalent bond formation,[225] as well as the chemical, physical, and kinetic properties of these derivatives. Resolution of racemic amino acid mixtures is accomplished by the hydrolysis of *N*-acetyl-DL-amino acids by the bound enzyme in a packed-bed reactor. The resulting mixture of the L-amino acid and the *N*-acetyl-D-amino acid is then treated by conventional means to give the desired crystalline L-amino acid. The process is fully continuous and includes the racemization of the unhydrolyzed *N*-acetyl-D-amino acid. At the moment, this process is the only documented example of a large-scale operation employing an immobilized enzyme. Others[164,323] have also immobilized aminoacylase and used it for the resolution of amino acid mixtures.

### *Group Addition Reactions*

There is only one good example reported of the use of an immobilized lyase for continuous conversions. Becker and Pfeil[354] immobilized D-oxynitrilase by adsorption on ECTEOLA-cellulose and used the conjugate in a packed-bed reactor for the continuous conversion of benzaldehyde to D-(+)-mandelonitrile (Equation 76). A good conversion (95% yield) and a product of good optical activity (97% D- and 3% L-mandelonitrile) were obtained. The method could also be used for the synthesis of other D-α-hydroxynitriles from corresponding aliphatic, aromatic, or heteroaromatic aldehydes. D-α-hydroxynitriles are important intermediates that can be readily transformed into optically active D-α-hydroxycarboxylic acids, substituted ethanolamines, or acyloins.

### *Isomerization Reactions*

Of all the isomerases immobilized, glucose isomerase, which converts glucose to fructose, seems to have the greatest economic potential. The enzyme has been immobilized by several means,[83,342,366] but the polyacrylamide lattice-entrapment preparation was described in greater detail. Strandberg and Smiley[366] isolated the enzyme from *Streptomyces phaeochromogenes,* immobilized it by gel entrapment, and examined its heat stability, pH-activity behavior, etc. Unfortunately, the enzyme (both free and immobilized) exhibited substantial inactivation and its use for long continuous operations seems questionable.

The use of immobilized ligases (synthetases) for synthetic purposes has not yet been reported. A member of this group has, however, been employed for separation purposes.[233](see page 143).

### *Analysis*

Immobilized enzymes have been used or proposed for various aspects and types of analysis. For discussion purposes, the different areas of analytical application may be divided into the following categories: (1) detection and determination of substrates and inhibitors of the enzymes, (2) determination of the microenvironment about polymers, (3) amino acid and peptide analysis, (4) nucleic acid analysis, and (5) structural analysis of macromolecules and multicomponent systems. The use of soluble and immobilized enzymes in analytical chemistry has been reviewed by Guilbault.[446-449]

### *Determination of Substrates and Inhibitors of the Enzymes*

Applications in this area range widely in sophistication. A rapid semiquantitative spot test for hydrogen peroxide was reported by Weetall and Weliky.[126] The test, which utilized horseradish peroxidase covalently bonded to CM-cellulose strips and which could be used to detect peroxide in small volumes at concentrations as low as 1 *μM,* was dependent on color formation with benzidine. The use of immobilized cholinesterase for determining the presence of certain insecticides and nerve gas type organophosphorus compounds has been advanced.[139,370,373] Guilbault and Kramer[373] developed a fluorometric assay for such anticholinesterase compounds with cholinesterase immobilized in starch gel pads. The assay was dependent on the hydrolysis of nonfluorescent substrates to fluorescent products by the active enzyme. Upon inhibition of the enzyme, the fluorescence dropped to a level approaching zero.

$$C_6H_5\text{–CHO} + HCN \xrightarrow{\text{D-Oxynitrilase}} C_6H_5\text{–CH(OH)–CN} \qquad (76)$$

D-(+)-Mandelonitrile

Guilbault and Das[370] later extended the work on cholinesterase and also reported on the determination of urea by immobilized urease. The concentration of the ammonium ions produced by the urease was measured either fluorometrically or electrochemically. Immobilized urease had been previously used by Riesel and Katchalski[46] for determining the urea level in serum and urine. Their analysis of the produced ammonium ions was based on a colorimetric method, and good agreement between their analyses and those obtained by the routine clinical method was observed. Weetall[145] has reported on immobilized sterol sulfatase for use in hydrolyzing steroid conjugates prior to analysis for total estrogens in body fluids.

Automated systems using immobilized enzymes have been developed for several purposes. Hicks and Updike[358] described an automated system for the determination of glucose and lactate using polyacrylamide-entrapped glucose oxidase and lactate dehydrogenase, respectively. The system contained a tiny packed bed of the enzyme, and substrate concentrations in each case were determined indirectly by spectrophotometric means using a coupled color-producing reaction (see Figure 39). A similar system employing polyacrylamide-immobilized glucose oxidase but using an oxygen electrode as the means for detecting changes in activities (and consequently substrate concentrations) was also reported.[386] A polarographic oxygen electrode measures the diffusion flow of oxygen through a membrane, and the current output of the electrode is a linear function of oxygen tension. According to Equation 77, the conversion of glucose by the immobilized glucose

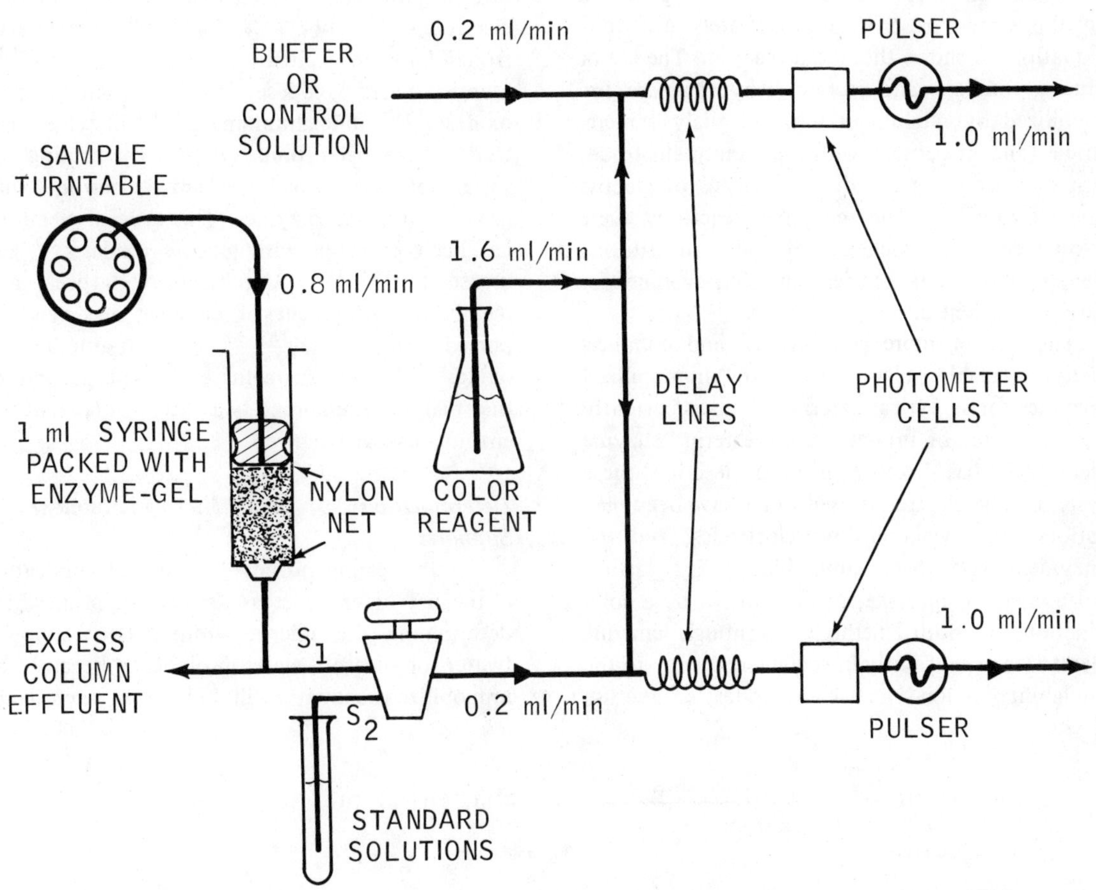


FIGURE 39. Diagram of instrumentation used for continuous analysis of glucose and lactate by immobilized glucose oxidase and lactate dehydrogenase, respectively. Enzyme column and tubing preceding column are at constant temperature. Steady state absorbance difference between the two photometer cells is a measure of the product concentration in the column effluent. A sample turntable changes the substrate solutions at a rate of 20/hr. (Redrawn from Hicks, G. P. and Updike, S. J., *Anal. Chem.*, 38, 726, 1966. With permission.[358])

oxidase causes a drop in the concentration and tension of oxygen that is detected by the oxygen electrode.

$$\beta\text{-D-glucose} + O_2 \xrightarrow{\text{Glucose oxidase}} \text{D-glucono-}\delta\text{-lactone} + H_2O_2 \quad (77)$$

Another automated system employing starch gel-immobilized cholinesterase was reported by Bauman et al.[372] The activity of the enzyme was monitored electrochemically using two platinum electrodes and an applied current of 2 $\mu$A. As long as the cholinesterase was active, the electrochemical system indicated a low potential (approximately 150 mV) because of oxidation of the thiocholine, the product of hydrolysis of butyrylthiocholine iodide, to the disulfide at the platinum anode. Conversely, as the activity of the immobilized cholinesterase decreased, the potential of the system rose to approximately 400 mV. Equation 78 shows the initial reaction. The use of glucose oxidase and lactate dehydrogenase immobilized by retainment within a dialysis membrane (and in contact with a platinum electrode) was likewise reported for the analysis of glucose and lactate.[436] The enzymic reactions were monitored electrochemically by substituting benzoquinone for oxygen and ferricyanide for ferricytochrome c.

The use of more conventional and commercially available electrodes with immobilized enzymes for analytical determinations is currently an active area of interest, and the term "enzyme electrode" has been coined to describe these systems. The electrodes used so far have been pH-, cation-, or oxygen-sensitive electrodes, and the enzymes have been immobilized by lattice-entrapment, membrane-entrapment, or by a combination of both methods. Although enzyme electrodes are currently in vogue largely due to the availability of ion-selective electrodes,[450] the use of enzyme-containing membranes coupled with hydrogen ion and oxygen-sensitive electrodes was reported in the early '60's by Clark and Lyons[438] for glucose oxidase and urease. The principle of the method in Figure 40 is for a cation-selective electrode employing L-amino acid oxidase immobilized by membrane entrapment (see Equation 79). The substrate, an L-amino acid, diffuses across the semipermeable membrane and is oxidized by the entrapped enzyme to an $\alpha$-keto acid, hydrogen peroxide, and an ammonium ion. The ammonium ions produced by the reaction are, in turn, sensed by the monovalent cation-selective electrode, and the potential of the electrode is measured after allowing sufficient time for the diffusion process to reach the steady state. The semipermeable membrane most commonly used in these enzyme electrodes is cellophane (dialysis tubing), and the enzyme electrodes based on this type of immobilization have been referred to as "liquid membrane electrodes."[383] Enzymes employed for such electrodes have been glucose oxidase,[438] urease,[438] L-amino acid oxidase,[383,437] D-amino acid oxidase,[388] and glutaminase.[390] Enzyme electrodes have been prepared also by coating the surface of oxygen- or ion-selective electrodes with lattice-entrapped enzymes. This type of electrode has been prepared with glucose oxidase[387] and urease.[384,385,392] Combination lattice- and membrane-entrapment of enzymes has been reported for urease,[384,385,393] D-amino acid oxidase,[388] and asparaginase.[388] The purpose of the dialysis membrane was to ensure that no enzyme leakage occurred.

### *Determination of the Microenvironment of Supports*

An interesting proposed analytical application is the use of enzymes as "molecular probes" for determining the microenvironment of supports (water-soluble or water-insoluble). Because an immobilized enzyme will behave in accordance

$$\underset{\text{Butyrylthiocholine}}{(CH_3)_3\overset{+}{N}CH_2CH_2S-\overset{O}{\overset{\|}{C}}-C_3H_7}\;I^- \xrightarrow{\text{Cholinesterase}} \underset{\text{Thiocholine}}{(CH_3)_3\overset{+}{N}CH_2CH_2SH}\;I^- + C_3H_7CO_2^- \quad (78)$$

$$R-CH(NH_2)-CO_2H + H_2O + O_2 \xrightarrow{\text{L-Amino acid oxidase}} R-C(=O)-CO_2H + H_2O_2 + NH_4^+ \quad (79)$$

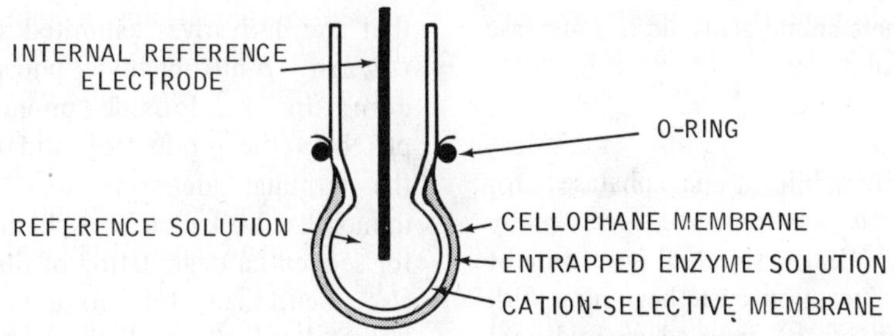


FIGURE 40. Cross-sectional view of liquid membrane enzyme electrode. The cellophane membrane retains the enzyme (L-amino acid oxidase) but permits diffusion of substrate (L-amino acid). Steady-state concentrations of product (ammonium ion) are monitored with cation-selective electrode.

with its local environment, a determination of specific enzymic properties (usually activity) can reveal the nature of its immediate environment. For example, an enzyme adsorbed or covalently bonded to a highly negatively charged support will exhibit a displaced pH-activity curve toward a more alkaline pH because of the higher concentration of hydrogen ions about the charged polymer and enzyme. Conversely, a positively charged carrier can induce an opposite effect. Consequently, because the behavior of an immobilized enzyme is dictated only by its local environment and not by the environment of the bulk phase (the phase that is usually measured by analytical means), it should be possible to determine the surface nature of supports with enzymes. Of course, it is assumed that no other effects are operating in producing the observed change. A comparison of the properties of an enzyme in different artificial environments with those of the enzyme in its native microenvironment (perhaps attached to subcellular particles or membranes) may well serve as a means for characterizing the native microenvironment itself. The use of this technique was recognized by McLaren[19] and by Goldstein, Levin, and Katchalski.[69] A more detailed discussion of the pH-activity behavior of adsorbed and covalently bonded enzymes can be found in the appropriate chapters on these enzymes conjugates.

### *Amino Acid Analysis*

The use of immobilized proteases as an alternative to conventional acid or base hydrolyses of proteins has been reported by several groups. This method offers an advantage over acid hydrolysis in that no destruction of serine, tyrosine, tryptophan, asparagine, and glutamine occurs. Royer and Green[216,217] immobilized pronase (a protease mixture of broad specificity from *Streptomyces griseus*) on the aminoaryl derivative of glass by diazotization and observed that the immobilized enzymes retained considerable activity toward large and small substrates. Although certain properties such as pH-activity behavior and temperature-dependence of the immobilized pronase were investigated, no firm comparison of the degradative ability of the enzyme conjugate with conventional hydrolyses was reported. The use of insoluble pronase as a tool in degradative studies had been suggested by Cresswell and Sanderson.[213] Recently, Bennett et al.[158] reported their findings on the degradation of peptides and proteins, using a mixture of individually immobilized prolidase, aminopeptidase-M, trypsin, and chymotrypsin bound to Sepharose. Incubations were conducted for 24 hr at 25°. Repeat digestions of synthetic β 1-24 adrenocorticotrophin revealed theoretical amino acid molar proportions with a standard deviation of less than 5% per residue. Quantitative release of amino acids was likewise observed for bradykinin and [5-valine]-angiotensin II amide. Incubations of two D-residue analogues of [5-valine]-angiotensin II amide, the [6-D-histidine]- and the [2-D-arginine]-angiotensin II amide, showed, however, that bonds adjacent to D-amino acid residues are not cleaved. Incubation of performic acid-oxidized ribonuclease revealed that aspartic acid and glutamic acid can be distinguished from asparagine and glutamine. Unfortunately, glutamine and methionine sulfone were not released quantitatively from this protein.

The amounts of these amino acids did not increase with incubation time.

### *Nucleic Acid Analysis*

The use of immobilized phosphatases for nucleic acid analysis was reported recently by Zingaro and Uziel.[39] In the standard procedure of RNA sequencing, a reaction cycle beginning with periodate oxidation of the exposed *vic*-hydroxyl group is used. Enzymatic removal of the newly formed phosphate monoester exposes the next *vic*-hydroxyl group and allows the next stage of the sequential degradation to begin (see Figure 41). If, however, the enzyme is not completely separated from the RNA, the procedure can lead to asynchrony by catalyzing the hydrolysis of the protective phosphate on the penultimate group during the oxidative elimination step. Consequently, sequential degradation of large nucleic acids is a major problem. The use of immobilized phosphatases to eliminate this difficulty was realized by Zingaro and Uziel,[39] who immobilized alkaline phosphatase by covalent bonding to various polymers (maleic anhydride-ethylene copolymer, crosslinked methacrylic acid-methacrylic acid-*m*-fluoroanilide copolymer, etc.) and showed that the derivatives exhibited substantial activity toward *p*-nitrophenyl phosphate, the four common nucleoside monophosphates, and ptRNAp (the product of oxidative elimination of the terminal adenosine of tRNA). The use of immobilized alkaline phosphatase in an apparatus for sequential degradation of ribonucleic acids was also mentioned, but no details were reported. Immobilized ribonuclease $T_1$[137] has also been suggested for use in structure studies of ribonucleic acids.

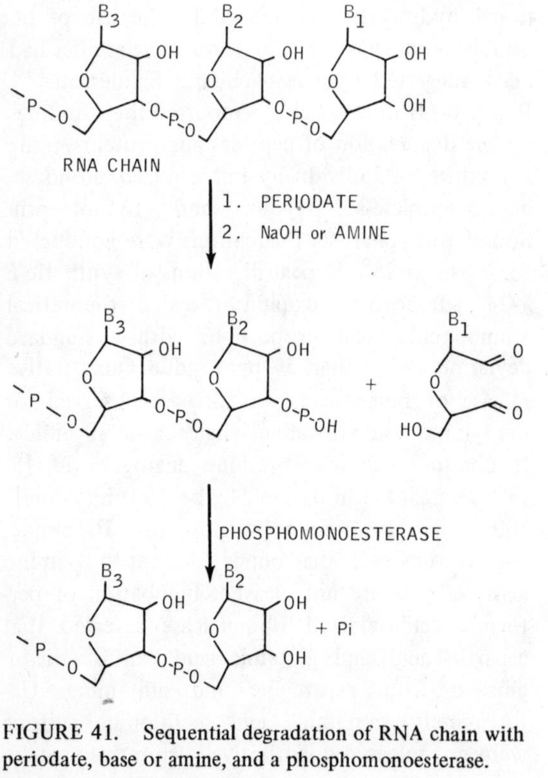

FIGURE 41. Sequential degradation of RNA chain with periodate, base or amine, and a phosphomonoesterase.

### *Structural Analysis of Macromolecules and Macromolecular Systems*

The complete degradation of macromolecules with various hydrolases followed by conventional analytical procedures gives the total composition of the material under investigation. On the contrary, the limited degradation of a macromolecule or a macromolecular assembly composed of different molecules can yield useful information about their structure. Immobilized enzymes are ideally suited for such purposes and have been used for various aspects of structural analysis of macromolecules. Only a representative portion of the reported work in this area is described below.

Limited or altered hydrolyses of proteins with immobilized proteases have been achieved (1) by reducing the contact time between protein and immobilized enzyme, (2) by a changed specificity induced in the enzyme upon immobilization, and (3) by a change in the solvent system. Time-dependent limited hydrolysis has been suggested and reported numerous times[52,166,169,172,196,198,199,202,218] using both continuous column and batch operations. A recent example of such a time-dependent limited degradation of a protein was reported by Gabel and v. Hofsten.[221] Degradation of apomyoglobin with an immobilized bacterial protease in a continuous column operation clearly revealed the presence of intermediate peptides as evidenced by either gel filtration chromatography or free zone electrophoresis of the digest. Repeated recycling of the effluents also showed progressive changes in the patterns reflecting additional degradation.

An excellent example of a change in the substrate specificity of an enzyme upon its immobilization was given by Ong et al.[173] A study of the comparative behavior of trypsin and immobilized trypsin (covalently bonded to a copolymer of maleic anhydride and ethylene) toward pepsinogen revealed that, whereas trypsin hydrolyzed 15 peptide bonds in the zymogen, the maximum number of bonds cleaved by the water-insoluble conjugate never exceeded 10. This cleavage pattern was independent of the ratio of carrier to enzyme, and the same difference was also observed if reduced carboxymethylated pepsinogen was used as the substrate. Different peptide maps of the two digests were also obtained. Such changed specificity (producing different peptide maps) of an enzyme after immobilization suggests the possibility of using the same enzyme on different supports as an approach for achieving overlapping or different peptide mapping regions in the same molecule. Alternatively and more conventionally, many soluble enzymes of different specificity can be used.

An interesting example of changed specificity in different solvents and a potentially highly useful application of immobilized trypsin for degradative purposes was reported by Gabel et al.[183] Trypsin, covalently bonded to cyanogen bromide-activated Sephadex G-200 and shown to retain its activity in 8 *M* urea, was used for the hydrolysis of lysozyme in 8 *M* urea. The behavior of lysozyme in urea is dependent on its unfolding by the denaturant and is contrary to that observed in the absence of urea; lysozyme in its native state is resistant to tryptic hydrolysis. Other water-insoluble derivatives of trypsin have also been shown to retain good activity in solutions of urea.[68,253]

The use of immobilized proteases for structural analysis has been applied to γ-globulin,[195-199] myosin and heavy meromyosin,[171,175,176,201,202] the 30S ribosomal proteins,[168] and bovine casein micelles.[281] Immobilized hyaluronidase[155] has been suggested for structural analysis of proteoglycans.

*Mechanistic Studies (Model Studies)*

Immobilized enzymes have been used for determining the nature of subunits of an enzyme, the chain folding of a protein, the mechanism of bioluminescence, and more generally as model systems for the natural membrane-bound enzymes. A few of the more interesting studies illustrating the potential of immobilized enzymes for mechanistic studies are discussed.

Certain steps in the sequence of reactions that result in the formation of a blood clot involve proteolytic enzymes. Figure 42 is a highly simplified scheme of the events that occur and is shown only to give the reader the location of particular enzymes or zymogens that have been immobilized. Only a few results of the extensive studies conducted by several groups are mentioned.

Thrombin has been immobilized by covalent attachment to various polymers[170,206-209] and the insoluble derivatives were shown to retain their clot forming ability,[206,207] to activate bovine factor V[208] (as did soluble thrombin), and to activate at an enhanced rate the zymogens of trypsin, chymotrypsin, and plasmin.[170] Water-insoluble thrombin was also reported to give the most controlled method of preparing cryofibrinogen – the cold insoluble fibrinogen.[209] Prothrombin[54,186,206] and its polytyrosyl derivative[54] were attached to the diazotized copolymer of *p*-aminophenylalanine and leucine. Immobilization of prothrombin did not affect the convertibility[54] or requirements of the activation.[54] As with the soluble zymogen, factors VII and X were crucial to biological and anion activation of the insoluble derivative. Likewise, trypsin converted insoluble prothrombin directly to thrombin without the requirement of additional factors. Water-insoluble trypsin derivatives[169,172] have also aided in the understanding of prothrombin activation. Polytyrosyltrypsin[172] bound to the copolymer of Phe and Leu activated prothrombin (with subsequent formation of thrombin) and factors VII and X present in the same zymogen preparation. Prothrombin was degraded by the water-insoluble trypsin conjugate of maleic anhydride and ethylene copolymer, but in this instance, no thrombin was detected. Furthermore, factor X, but not factor VII, was activated by this derivative. The difference in behavior of the two trypsin conjugates with regard to thrombin formation was attributed to the fact that under the experimental conditions used, thrombin was readily degraded by the maleic anhydride-ethylene copolymer conjugate; the polytyrosyl derivative of trypsin (or even native trypsin) did not degrade thrombin. Plasminogen[210] and streptokinase[218,220] have been immobilized. Insoluble plasminogen,[210] prepared by covalent coupling to the previously mentioned copolymer of Phe and Leu, was found to be active as a proactivator but not as a plasminogen; i.e., on the addition of streptokinase, activator but not plasmin was formed. Immobilized streptokinase, bound to either the diazotized copolymer of Phe

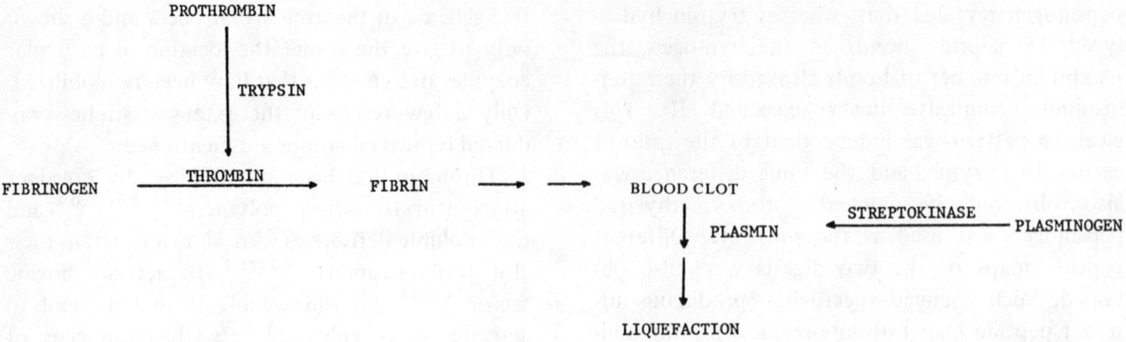


FIGURE 42. Partial sequence of reactions in formation and dissolution of blood clot. Scheme shows proteolytic enzymes that have been immobilized.

and Leu[218] or to *p*-aminobenzylcellulose,[220] converted plasminogen to plasmin.

The use of water-insoluble proteases for the controlled activation of other zymogens has been reported. For example, water-insoluble trypsin[57] was used to effect a controlled chymotrypsinogen to chymotrypsin conversion. After the conversion, the water-insoluble trypsin was easily removed and caused no contamination difficulties. Pepsinogen[159,173] was likewise activated by water-insoluble trypsin. A water-insoluble derivative of renin[185] has been shown to be active in angiotensin formation from angiotensinogen.

Water-insoluble enzyme conjugates have been proposed as model systems for studying the behavior of natural membrane-bound enzymes, and a rather large number of water-insoluble enzymes have been prepared. The comparative behavior of natural membrane-bound, "solubilized," and synthetically immobilized enzymes has also been reported. The rationale is to obtain information on the possible influence of the support on the characteristics of the enzyme. Brown, Chattopadhyay, and Patel[50] examined the comparative behavior of membrane-bound, digitonin-solubilized and CM-cellulose azide-linked sarcoplasmic recticulum ATPase isolated from rabbit hearts and observed that there were differences in the pH optimum, the ouabain sensitivity, and the metal ion stimulation of the three preparations. The membrane-bound ATPase exhibited an activity maximum at pH 8.1; the solubilized and matrix-supported preparations had activity maxima at pH 7.2. The membrane-bound ATPase was inhibited by 10 m*M* ouabain and stimulated by 1 m*M* $MgCl_2$ and endogenous NaCl and KCl at 1 m*M* and 2 m*M*, respectively. The solubilized enzyme, however, was slightly stimulated by 10 m*M* ouabain and was only moderately stimulated by 1 m*M* $MgCl_2$. Solubilized ATPase was inhibited by 0.1 *M* NaCl and KCl. The matrix-bound ATPase was inhibited by 10 m*M* ouabain (similar to the membrane-bound enzyme) and was stimulated by 1 m*M* $MgCl_2$. The cellulose-supported ATPase was, however, also inhibited by 0.1 *M* NaCl and KCl (similar to the solubilized enzyme). Brown et al.[50] attributed the difference in behavior of the three preparations to conformational differences. The membrane-bound enzyme has a conformation that results in sensitivity to ouabain and to metal ions. Upon solubilization, the natural conformation of the ATPase is lost and hence a drastic change in sensitivity occurs. Upon reattachment of the solubilized enzyme to a water-insoluble support, the required conformation for ouabain inhibition is restored. The failure to restore the original $Na^+$ and $K^+$ ion sensitivity of the enzyme upon its immobilization was ascribed to incomplete restoration of the required conformation. Further examples of such comparative studies of membrane-bound systems have been reported by Brown and his colleagues.[51,226,227] Whittam, Edwards, and Wheeler[98,230] have likewise examined immobilized apyrase and ATPase systems.

A recent brief communication by Chan,[232] using the concept of immobilization for studying the function of enzyme subunits and the effect of subunit interaction, is an exceedingly good example of the application of immobilized enzymes for mechanistic studies. Chan's method consists of covalently attaching a subunit-composed enzyme to a *low level functionalized* polymer, dissociating the individual subunits, and finally removing the noncovalently bonded units from the polymer-bound subunits (see Figure 43).

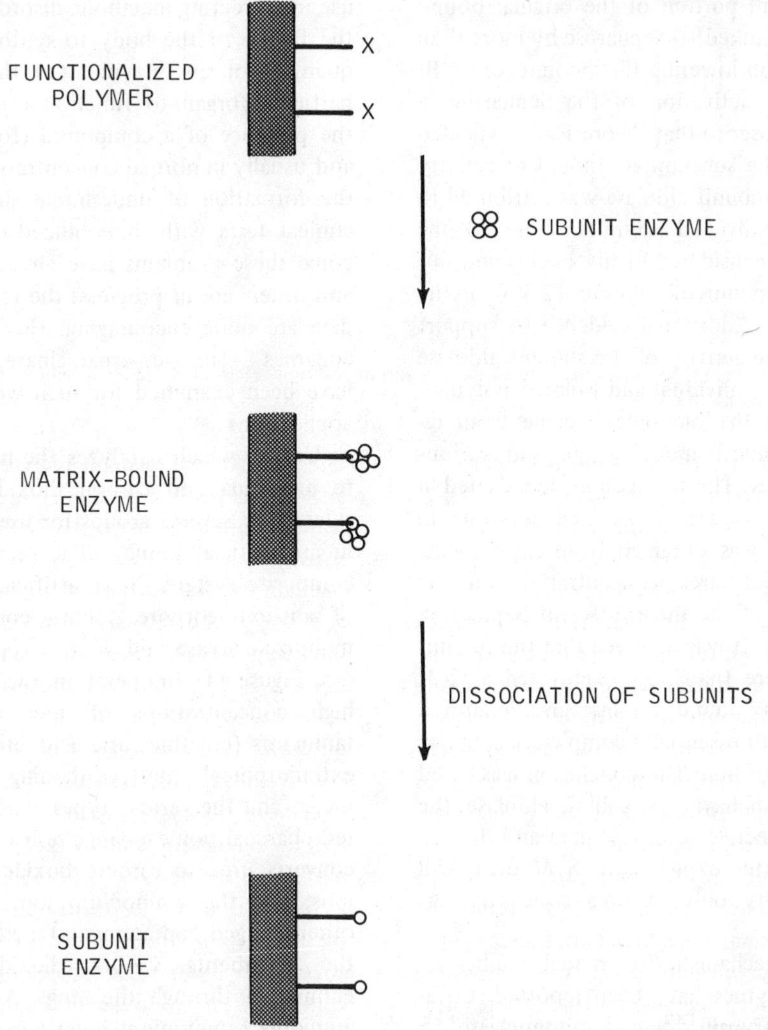


FIGURE 43. Steps involved in the investigation of subunits of an enzyme via the immobilization approach.

In effect, individual subunits are retained on the polymer, and association is prevented by the rigidity of the support. This method offers an opportunity to study the nature of an individual subunit or the subunit interactions of an enzyme. Low level functionalization of the support is necessary in order to ensure that only individual subunits are attached and that these are spatially distant from each other. Chan tested the validity of this approach with the tetrameric subunit enzyme aldolase. The enzyme was attached to cyanogen bromide-activated Sepharose, and dissociation of the subunits was achieved by 8 *M* urea solutions containing mercaptoethanol. An examination of the activity and protein content of the matrix-bound aldolase (the Sepharose-bound conjugate washed only with 1 *M* NaCl phosphate buffer), the subunit aldolase (the conjugate after the 8 *M* urea wash), and matrix-bound *renatured* aldolase (prepared by adding back excess aldolase subunits to the subunit aldolase in 8 *M* urea before dialysis) revealed that each derivative exhibited activity (4.5, 1.6, and 2.2 mU/μg, respectively) and that the protein content of the subunit aldolase was 27.5% that of the matrix-bound aldolase. A protein content of 25% in the subunit aldolase would be expected on the basis of the change from a tetrameric to a monomeric state. According to Chan, the difference between the actual and theoretical values suggested the possibility that a small portion of the original bound enzyme may be linked to Sepharose by more than one subunit. Upon lowering the amount of CNBr used during the activation of the Sepharose, a value (24.8%) closer to that theoretically expected was obtained. The substantial amount of activity observed in the subunit aldolase was attributed to the activity of individual subunits of the enzyme (or at least was considered highly likely) and not to the excess amount of subunits (2.5%) of the subunit aldolase. Additional evidence to support the claim that the activity of the subunit aldolase was due to only individual and isolated polymer-bound subunits (the monomers) came from re-association, chemical modification, and various denaturant studies. The strongest evidence cited in favor of the existence of isolated subunits in subunit aldolase was obtained from experiments on the effect of urea concentration on the enzymic activity of the three different Sepharose-bound conjugates. It was observed that the subunit aldolase was more than 50% inactivated at 2 *M* urea while matrix-bound aldolase and renatured aldolase were both essentially completely active at this urea concentration. This conclusion was based on the reported behavior of soluble aldolase; the enzyme is fully active in 2.3 *M* urea and dissociated aldolase (after exposure to 8 *M* urea) will regenerate activity only at urea concentrations lower than 2.3 *M*.

Additional mechanistic or model studies of immobilized enzymes have been reported. Covalently bonded trypsin[132] and ribonuclease[132] were employed for investigating the nature of protein folding, luciferase[118] was used for examining the mechanism of bacterial luminescence, and immobilized trypsin[45] was utilized for determining the retention of enzymic activity in an enzyme (trypsin) upon proteolysis by another enzyme (pronase). Further, polyacrylamide-entrapped peroxidase[389] and alkaline phosphatase[359] conjugates were employed as model systems for cytochemical staining reactions. The use of Sepharose-bound tRNA nucleotidyltransferase[135] for studying interactions between the enzyme and tRNA or tRNA-like structures was proposed.

*Metabolic Disorders*

Immobilized enzymes have been suggested for use in correcting metabolic disorders caused by (1) the failure of the body to synthesize a sufficient quantity of an enzyme, (2) the inability of a particular organ to function effectively, and (3) the presence of a compound (found in the body and usually in normal concentrations) that induces the formation of undesirable side effects. Some clinical tests with immobilized enzymes to overcome these problems have already been reported and others are in progress; the results obtained to date are quite encouraging. However, only a few enzymes – urease, asparaginase, and catalase – have been examined for such worthwhile clinical applications.

Urease, which catalyzes the hydrolysis of urea to ammonia and carbon dioxide, is being considered by several groups for use as a component in an artificial kidney. The presently envisioned composite system for an artificial kidney consists of an extracorporeal shunt containing the immobilized urease and various types of adsorbents (see Figure 44). In operation, the blood containing high concentrations of urea and other contaminants (creatine, uric acid, etc.[403]) enters the extracorporeal shunt containing the immobilized urease and the various types of adsorbents (activated charcoal, ion-exchange resins, etc.). The urease converts urea to carbon dioxide and ammonium ions, and the ammonium ions (along with the other charged contaminants) are then adsorbed on the adsorbents. Carbon dioxide is eventually eliminated through the lungs. Adsorption of the ammonia (ammonium ions) is necessary in order to eliminate the more serious problem associated with high concentrations of this ion. The adsorbents are also necessary in order to remove the other charged and unwanted contaminants of the venous blood. Major contributions to the use of immobilized urease for this purpose (or that of urea removal in general) have been made by Chang, Sparks, Gardner, and their many co-workers. A brief survey of their results follows.

Chang[394,395,398] first immobilized urease by microencapsulation and reported that the encapsulated urease, injected intraperitoneally, increased markedly the concentration of ammonia in the blood of a dog. In subsequent studies, Chang[401] described an extracorporeal shunt chamber and proposed its use, loaded only with microencapsulated urease, for an artificial kidney. In dogs, the systemic blood ammonia levels were regularly increased when the shunt chamber con-

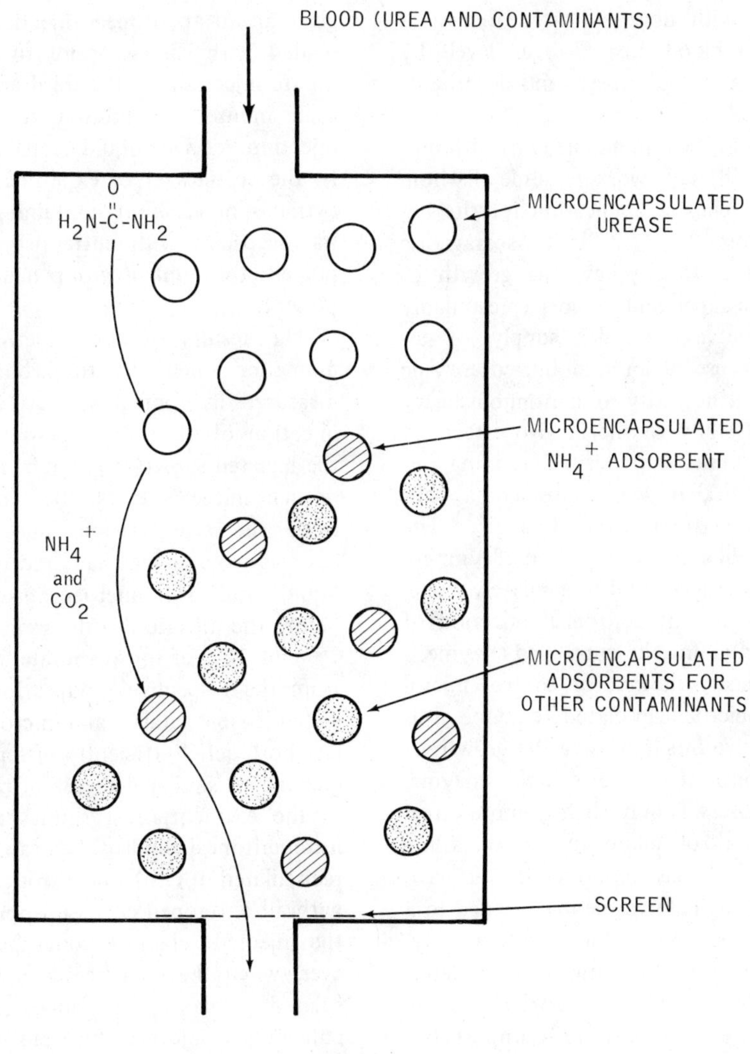


FIGURE 44. Artificial kidney based on immobilized urease. Extracorporeal shunt contains microencapsulated enzyme and adsorbents.

tained urease-loaded microcapsules. The use of microencapsulated adsorbents (likewise encapsulated to prevent contamination and immunological reactions) was also investigated. Further studies by Chang and co-workers dealt with the preparation of nonthrombogenic microcapsules,[399] animal[399,404] and human[403,415] experiments employing extracorporeal shunts (with and without microencapsulated urease), and the use of microencapsulated activated carbon.[403,415]

Sparks et al.[408] also reported the use of microencapsulated urease and various adsorbents in an extracorporeal chamber for employment as an artificial kidney. The use of transition metal complexes on Dowex® A-1 for ammonia adsorption was investigated.

Levine and LaCourse[407] presented a design theory for a compact artificial kidney utilizing microcapsules containing urease, ion-exchange resin, and adsorbents for uric acid and creatine, and showed that the critical factor was the permeability of the microcapsule wall. Calculations revealed that an artificial kidney 10 cm in length and 2 cm in diameter was feasible.

Uremic detoxification by oral ingestion of a mixture of microcapsules (containing urease, activated charcoal, and ion-exchange materials) was reported recently by Gardner et al.[255,256] Preliminary studies with dogs indicated that it is possible to lower blood urea nitrogen levels by feeding microencapsulated urease and adsorbents to azotemic animals.

Asparaginase has been immobilized by different methods and by different workers, some of whom have suggested a clinical application for the immobilized enzyme.[103,222,410] L-Asparaginase has been observed to suppress the growth of certain asparagine-dependent tumors (presumably by depleting the extracellular supply of asparagine) and the use of an immobilized enzyme could eliminate the necessity for continuous heavy dosages. A good study of the in vivo effects of semipermeable microcapsules containing L-asparaginase on 6C3HED lymphosarcoma cells implanted in mice was reported by Chang.[410] The mice received a subcutaneous injection of lymphosarcoma cells in the groin and immediately thereafter were given an intraperitoneal injection of microcapsules containing the entrapped enzyme, a soluble asparaginase solution, or control microcapsules. The microencapsulated enzyme was much more effective in suppressing the growth of the lymphosarcoma than the soluble enzyme. Soluble asparaginase was only slightly more effective than the control microcapsules or saline solution. Weetall[103] has reported the in vivo stability of L-asparaginase covalently bonded to a Dacron vascular prosthesis in dogs; unfortunately, no asparagine levels in the animal were determined. Hasselberger et al.[222] reported the immobilization of the enzyme on various supports but only suggested its use in cancer therapy; no in vivo experimental work was described.

Acatalasaemia is a rare inherited metabolic disease of man and animals caused by a deficiency of catalase. Symptoms of the disease in man are oral sepsis followed by severe ulcerative gangrenous lesions of the oral cavity. Recently, Chang and Poznansky[402] have suggested the use of microencapsulated catalase for treating this disease and have investigated the possibility in acatalasaemic mice. Two types of in vivo experiments were conducted. In one set of experiments, one group of normal mice and three groups of acatalasaemic mice (a mutant strain) were used. Each mouse received a subcutaneous injection of sodium perborate (a substrate precursor). As controls, one group of acatalasaemic and the group of normal mice received no other treatment. In the second group of acatalasaemic mice, each animal was given an intraperitoneal injection of microcapsules loaded with catalase prior to receiving the perborate injection. In the third acatalasaemic group, each mouse, in addition to the subcutaneous injection, received liquid catalase intraperitoneally. In the second set of experiments, the use of an extracorporeal shunt containing a suspension of microcapsules with entrapped catalase was employed; the control group had no microcapsules present.

The results of the experiments were quite dramatic. When the total body perborate was measured in control animals 20 min after the injection of the sodium perborate, only 2.5% of the injected substrate could be recovered from the normal mice, whereas 70% could be recovered from the acatalasaemic mice. Yet when the acatalasaemic mice had received injections of liquid catalase or microencapsulated catalase just before the injection of the sodium perborate, only 7% and 16% of the perborate could be recovered from the respective groups. Intraperitoneally injected soluble catalase and microencapsulated catalase both acted efficiently in replacing the missing enzyme. A similar decrease of perborate was seen in the extracorporeal shunt experiments. When intraperitoneal fluid of the acatalasaemic mice was recirculated for 20 min through shunt chambers without entrapped catalase, approximately 76% of the injected substrate could be recovered. However, when the chambers contained microencapsulated catalase, only about 32% of the subcutaneously injected substrate was recovered. No enzyme leakage from the microcapsules (or breakage) was found within the limits of experimental detection.

The use of bound tyrosinase[122] for the in vivo and in situ synthesis of L-DOPA and of surface-entrapped proteases[362] for the prevention of blood clots are other suggested clinical applications of immobilized enzymes.

*Fuel Cells*

The use of immobilized enzymes for enzyme-dependent fuel cells has been investigated, but only to a rather limited extent. Young et al.[451] discussed the various types of microbial-based fuel cells (depolarization, product and redox cells) and the theoretical limitations of each type in terms of the metabolic activities of microorganisms and the coupling of these to electrochemical systems. Although the use of cell extracts and enzymes, particularly in an insoluble form, was considered, no specific enzyme or system was described. Blumenthal et al.[439] did describe an enzymatic fuel cell based on the hydrolysis by immobilized papain of the uncharged substrate, *N*-acetyl-L-glutamic acid diamide to ammonium *N*-acetyl-L-glutamine. Immobilization of papain was achieved by retainment within a compartment whose sides were composed of semipermeable membranes (cation or anion-exchange membranes). The fuel cell consisted of two 50 ml electrode compartments $P_1$ and $P_2$ (for measuring potentials), two 100 ml compartments I and II in which Teflon®-coated magnetic stirrers were inserted, and one 10 ml inner compartment, i, which contained the enzyme. The complete system of Blumenthal et al. is shown in Figure 45. Two sets of membranes with different properties were used for the composite inner structure. Both sets contained the

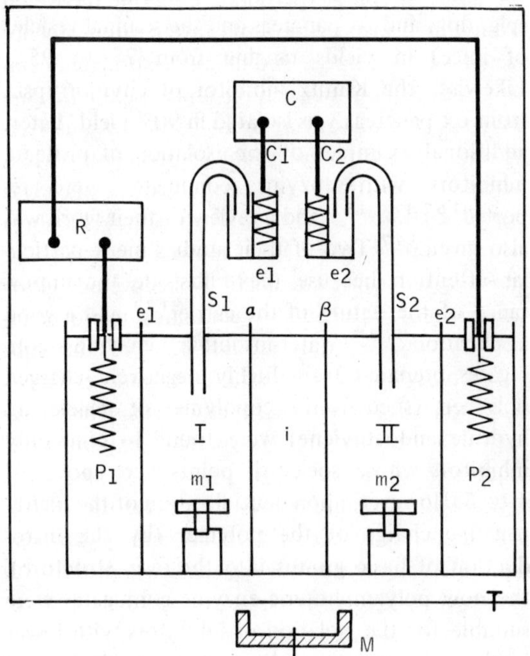


FIGURE 45. Enzymatic fuel cell based on membrane entrapped enzyme. Potential-measuring half-cells ($P_1$ and $P_2$); current-carrying half-cells ($C_1$ and $C_2$); cation-exchange membranes ($e_1$, $e_2$, and $\beta$); anion-exchange membrane ($\alpha$); magnetic stirrers ($m_1$ and $m_2$); salt bridges ($S_1$ and $S_2$); potential recorder (R); constant-current source (C); air thermostat (T); driving magnet (M); outer 100 ml compartments (I and II); inner 10 ml compartment (i). Concentration of papain, 1 mg/ml. Membrane area, 20 $cm^2$. (Redrawn from Blumenthal, R., Caplan, S. R., and Kedem, O., *Biophys. J.*, 7, 735, 1967. With permission.[439])

same membrane $\beta$ (a cation-exchange membrane equilibrated with ammonium ions), but differed in the membrane $\alpha$. In one set, an anion-exchange membrane, $\alpha_I$, was used; the other set contained an anion-exchange membrane, $\alpha_{II}$, which had been treated with strong NaOH for 24 hr and had lost most of its charged groups. Both membranes were equilibrated with the *N*-acetyl glutamine ion as the negative counter-ion. The membrane area was 20 $cm^2$. This composite membrane system gave rise to an electromotive force, ultimately reaching a steady state, when positioned between two identical solutions in which the affinity of the reaction had been fixed. Further, calculations revealed that a very high degree of coupling was necessary for energy conversions with reasonable efficiency. This, in turn, meant the necessity for highly permselective membranes. The values for maximum efficiency of energy conversion, calculated from measured values of the degree of coupling for the two sets of membranes, were 25 and 7% for the $\alpha_I$ and $\alpha_{II}$ set, respectively.

Wingard et al.[380] examined immobilized glucose oxidase for fuel cell purposes. The enzyme was immobilized by lattice entrapment within polyacrylamide gel and was employed in an enzyme-electrode fashion using a platinum electrode. Constant current voltametry was used for the evaluation of design variables. The half-cell potential for the oxidation of the substrate at constant pH was measured, and the potential was a linear function of the current. The open-circuit half-cell potential at zero current could be obtained by extrapolation. Differences between the measured potential and the open-circuit potential at any constant current were indicative of the loss of useful voltage due to concentration gradients within or adjacent to the enzyme electrode assembly. A serious concern was expressed by these investigators for the applications of the system as a fuel cell because of the high degree of concentration polarization observed.

## *Facilitated Transport*

The use of immobilized enzymes for facilitated transport was demonstrated recently by Broun et al.[257] Carbonic anhydrase was immobilized on hydrophobic sheets of Silastic by intermolecular crosslinking with glutaraldehyde. Preparation of the membrane-enzyme systems consisted of gently spreading a mixture of the crosslinking reagent and carbonic anhydrase on the Silastic sheets (on one or both sides of the membrane), allowing the crosslinking to occur at 4°, and then washing the layered membrane systems until free of glutaraldehyde and soluble enzyme. A Silastic membrane was employed in this study because carbon dioxide can cross the hydrophobic film, but water and ions such as bicarbonate cannot permeate. The extent of diffusion of carbon dioxide through the stretched enzyme-containing films (stretched between two 250 ml chambers) was made by measuring the transported $CO_2$ either by tracer or pH-stat techniques. Results of measurements of $CO_2$ transport revealed that when carbonic anhydrase is bound to the $CO_2$ donor side of the membrane, the transport of $CO_2$ was increased 1.5 times compared with unmodified Silastic sheets. Further, when the enzyme is bound to both sides of the membrane, the $CO_2$ transport is increased by a factor of 2. The significance of facilitated transport in biological systems was discussed and later amplified by these workers.[250]

**Applications Dependent on Structural Properties of Enzymes**

*Separations*

Immobilized enzymes have been used in applications that are dependent only on the structural aspects of the protein. In the specific examples to be discussed, the purpose of the immobilized enzyme was to provide a highly specific adsorbent (in water-insoluble form) for the separation and purification of another component. Only the enzyme's characteristic property of specificity was used. Water-insoluble enzyme conjugates have been employed for the isolation of natural enzyme protein inhibitors, antibodies, ribonucleic acids, and active-site fragments of degraded enzymes; immobilized zymogens have been used for the resolution of racemic amino acid mixtures.

The usual procedure for the isolation of an inhibitor, ribonucleic acid, or other component of a mixture consists of preparing a water-insoluble conjugate of the enzyme, contacting the immobilized protein with the mixture containing the enzyme-specific component under conditions where the two have the maximum but reversible degree of interaction (binding), washing the immobilized enzyme-component complex to remove all other nonspecifically held contaminants, and finally eluting (desorbing) the desired component from the water-insoluble derivative. Optimal conditions for desorbing a particular component from the water-insoluble enzyme complex must be experimentally determined but often can be achieved by a change in the pH, ionic strength, or temperature of the medium or through the addition of soluble substrates or inhibitors of the enzyme. Both batch and column operations may be employed for such separation purposes. The separation of various components on other than water-insoluble enzyme conjugates is mentioned briefly in Chapter 11.

Most separation applications of immobilized enzymes have been made with water-insoluble derivatives of the hydrolases, and the major portion of these studies has been conducted by Fritz and his co-workers.[163,167,174,177,179,180,181] Fritz et al.[174] prepared water-insoluble conjugates of trypsin, α-chymotrypsin, and kallikrein by covalent bond formation to the maleic anhydride-ethylene copolymer. Trypsin inhibitors were isolated from crude organ extracts (pig, dog, and ox pancreas and the seminal vesicles of mice) in yields ranging from 75 to 95%. Likewise, the Kunitz inhibitor of chymotrypsin from ox pancreas was isolated in 90% yield. Later, additional examples of the isolation of protease inhibitors with enzyme conjugates were reported[167,180,181] and a review of their work was also given.[177] Two of their studies merit particular attention because they illustrate the importance of the nature of the carrier[179] and a good application.[163] Water-insoluble enzyme conjugates prepared from highly negatively charged polymers (such as the copolymer of maleic anhydride and ethylene) were found to bind only inhibitors whose isoelectric points were above pH 4 to 5. However, upon neutralization of the highly negative charge of the polymer (by the introduction of basic groups into the resin structure), the now polyamphoteric enzyme conjugates were suitable for the isolation of inhibitors with lower isoelectric points. In addition, the polyamphoteric conjugates possessed a higher specific binding capacity for inhibitors than the corresponding polyanionic enzyme conjugates; they also required a much lower hydrogen ion concentration for elution. Charge neutralization was achieved by the addition of *N,N*-dimethylethylenediamine to the negatively charged polymer after sufficient enzyme coupling had taken place.

The use of a water-insoluble trypsin – trypsin-kallikrein inhibitor[163] conjugate for the preparation of modified protease inhibitors once again illustrates the potential of water-insoluble derivatives. Modified derivatives of the trypsin-kallikrein inhibitor from bovine organs were prepared by adding either maleic anhydride or pyridoxal-5′-phosphoric acid to a suspension of the water-insoluble trypsin-inhibitor conjugate and then removing the excess reagents and products from the resulting modified insoluble complexes by washing with buffer solutions. The complexes were dissociated from the resins by slightly acidic buffer solutions, and the modified inhibitors (the tetramaleoyl and poly (5′-phosphopyridoxyl derivatives) were eluted and isolated.

The use of water-insoluble trypsin[272,278] and α-chymotrypsin[192,272] for the isolation of inhibitors has also been reported by others.

Immobilized enzymes have been employed for the isolation of antibodies,[143,249,272] ribonucleic acids,[233] and affinity-labeled active-site fragments[133] of enzymes. Water-insoluble peroxidase[272] was used for the purification of Jack bean agglutin, glutamic-aspartic transaminase[294] for the isolation of its antienzyme from rabbit antiserum, and staphylococcal nuclease[143] for the isolation of its antibody from the γ-globulin fraction of rabbit antiserum. Sepharose-bound isoleucyl-tRNA synthetase[233] from *E. coli* was used to isolate isoleucyl-tRNA from crude tRNA preparations.

The isolation of affinity-labeled fragments of an enzyme through the use of an immobilized enzyme (the same enzyme) reported by Givol et al.[133] warrants further discussion (see Figure 46). Givol and co-workers reacted ribonuclease with the diazonium derivative of 5′-(4-aminophenyl phosphoryl) uridine 2′(3′)-phosphate (PUDP) – an affinity label compound for the enzyme – and obtained a derivatized RNase labeled specifically at a tyrosyl residue. This derivatized RNase was reduced, carboxymethylated in 8 *M* urea, and the alkylated protein was digested with trypsin. The

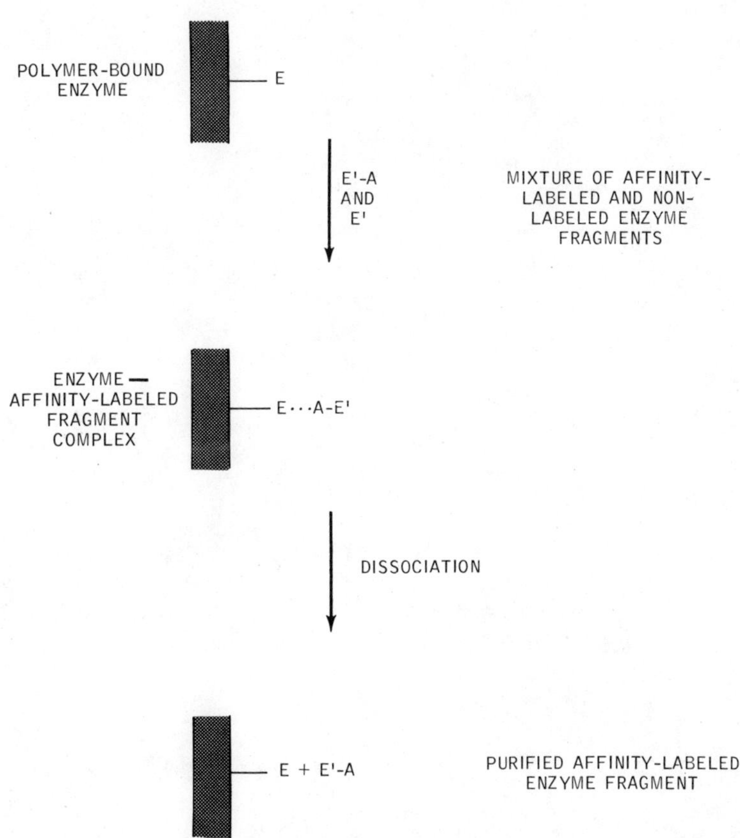


FIGURE 46. Isolation of affinity-labeled peptide from the hydrolyzed mixture of an enzyme by the immobilized enzyme.

digest was passed through a column containing immobilized RNase (covalently bonded to Sepharose) and eluant fractions were collected. Absorbance measurements of these fractions indicated that no peptide labeled with PUDP emerged from the column; most of the digest passed through. Upon washing the column with water and sodium acetate buffers, some additional peptides were removed but these again did not contain any PUDP ligand. Finally, the yellow PUDP peptide still bound to the Sepharose-RNase column was eluted as a sharp band with 0.8 *M* ammonium hydroxide. Analysis of this PUDP-peptide (after one paper electrophoresis) showed it to be identical to a PUDP-peptide of RNase A isolated previously by several steps of chromatography and electrophoresis. The successful use of this approach for the isolation of an affinity-labeled peptide from RNase should encourage more attempts.

Finally, the resolution of amino acid derivatives using covalently bonded trypsinogen[187] has been briefly reported.

Chapter 11

# RELATED TOPICS

## Immobilized Coenzymes and Coenzyme Analogs

Many of the enzymes immobilized to date, and an even larger number that would be desirable to have immobilized, require coenzymes for catalysis. Coenzymes are expensive and labile compounds, and it is essential to eliminate their continuous addition if coenzyme-dependent enzymes are to be used efficiently. One approach for doing this, first advanced by Larsson and Mosbach[452] and subsequently by others,[453] involves the covalent attachment of these water-soluble, low molecular weight compounds to water-insoluble matrices.

Larsson and Mosbach[452] immobilized nicotinamide adenine dinucleotide ($NAD^+$) by covalently attaching the oxidized coenzyme to a previously prepared Sepharose-ε-amino caproic acid conjugate via DCC.

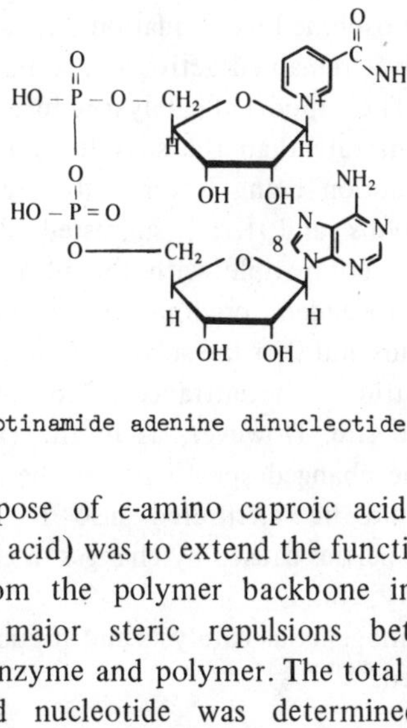

Nicotinamide adenine dinucleotide (NAD)

The purpose of ε-amino caproic acid (6-aminohexanoic acid) was to extend the functional group away from the polymer backbone in order to prevent major steric repulsions between the soluble enzyme and polymer. The total amount of gel-bound nucleotide was determined spectrophotometrically (after HCl hydrolysis) and found to be 56 μmoles/g of dry polymer, which corresponded to a maximum binding yield of approximately 20%. A control, to which no DCC had been added, retained 11 μmoles/g of polymer, but repeated washings did not remove this material. The control nucleotide-Sepharose conjugate did not, however, show the same properties of $NAD^+$ as the DCC-prepared material. The properties exhibited by the carbodiimide-coupled coenzyme (but not by the control) were (1) the absorption maxima at 340 and 260 nm upon dithionite reduction, (2) the absorption maxima at 325 and 260 nm of the CN-addition compound, and (3) the normal coenzymic action toward soluble lactate dehydrogenase in the oxidation of lactate to pyruvate (Equation 80). Larsson and Mosbach reasoned from the comparative behavior of these two materials that the control contained, at most, only negligible amounts of material with coenzymic function and that the control Sepharose-nucleotide conjugate, as determined by its absorption at 260 nm, was not $NAD^+$.

The bound $NAD^+$ showed approximately 0.2% of coenzymic function as compared with soluble $NAD^+$. Reversibility of the coenzymic function (i.e., the formation of lactate from pyruvate) was also observed with lactate dehydrogenase. The bound $NAD^+$ possessed good storage stability; it was still active after several weeks in water at 4° and could be used repeatedly without loss of activity.

Neither the reason for the rather low coenzymic activity nor the precise mode of binding of the $NAD^+$ to the matrix was established. Low activity could be attributed to steric hindrance by the gel matrix or to some electronic or conformational effect of the derivatized $NAD^+$. It was suggested that binding occurred through ester formation between the hydroxyl groups of the ribose moieties of the pyridine nucleotide and the carboxyl groups of the Sepharose-ε-amino caproic acid conjugate.

Weibel, Weetall, and Bright[453] immobilized $NAD^+$ by covalent attachment to porous and solid glass particles via the diazotized aminoaryl derivative. The bound coenzyme exhibited coenzymic activity toward yeast alcohol dehydrogenase, and the activity was reversible (Equation 81). Solid glass beads were found to be superior to porous beads because they did not strongly adsorb

$$CH_3CHOHCO_2^- + NAD^+ \xrightleftharpoons{\text{Lactate dehydrogenase}} CH_3COCO_2^- + NADH + H^+ \quad (80)$$

L-Lactate                Pyruvate

$$CH_3CH_2OH + NAD^+ \xrightleftharpoons{\text{Alcohol dehydrogenase}} CH_3CHO + NADH + H^+ \quad (81)$$

the coenzyme. In the case of the porous beads, small but measurable amounts of $NAD^+$ were slowly released even after thorough washings. As with $NAD^+$ binding to Sepharose, the exact position of coupling of the coenzyme molecule was not established; it was most likely at the C-8 position of the adenine nucleus.

Others have reported the immobilization of coenzymes and coenzyme analogs to various matrices for purposes such as the isolation of coenzyme-dependent enzymes or mechanistic studies. The isolation of coenzyme-dependent enzymes with the use of covalently bonded, water-insoluble coenzymes is dependent on the strength of the binding between the two and is an example of what is presently referred to as "affinity chromatography." Larsson and Mosbach[452] and Weibel et al.[453] also suggested the potential use of their immobilized $NAD^+$ preparations for this purpose.

Lowe and Dean[454] in a recent publication reported the covalent attachment of $NAD^+$ and $NADP^+$ directly to cellulose via the cyanogen bromide activation procedure and indirectly to cellulose derivatives of ε-amino caproate, glycylglycylglycine, and *o*-aminophenol. The coenzymes were coupled to ε-aminocaproyl- and glycylglycylglycyl-cellulose through the carbodiimide procedure using CMC and to *o*-hydroxyanilino-cellulose via treatment of this derivative with diazotized benzidine. Only $NAD^+$ was coupled by the diazotization method. The bound coenzymes were used for the resolution of various dehydrogenases. Lowe and Dean made a most interesting and significant discovery concerning the binding behavior of dehydrogenases that could also be relevant to the question of the low coenzymic activity observed by Larsson and Mosbach.[452] Their finding was that the conformation of the bound coenzyme has a more profound effect on binding of the enzyme than the distance of the cofactor from the backbone of the matrix. Unfortunately, Lowe and Dean did not test the coenzymic activities of their preparations.

Other examples of the covalent attachment of coenzymes or analogs to water-insoluble supports are $N^6$-(6-aminohexyl)-AMP[455] to Sepharose, $N^6$-ε-aminocaproyl-cyclic-AMP[456] to Sepharose and AMP[457] to chloromethylated polystyrene. The coenzymes FMN[458] and FAD have been entrapped in methylcellulose films. $NAD^+$[440] was also covalently bonded to a water-soluble polymer.

## Immobilized Homogeneous Catalysts

Although far removed from the realm of biochemistry, it is interesting to note that recently the concept of covalently bonded enzyme immobilization has been applied to other soluble (homogeneous) catalysts. For example, Grubbs and Kroll[459,460] reported the attachment of a rhodium(I) hydrogenation catalyst to a chloromethylated-divinylbenzene-styrene copolymer through a phosphine link (Equation 82).

The catalyst remained active and could be used repeatedly. The supported catalyst reduced olefins at a different rate than the soluble catalyst, the rate of reduction being lower with larger sized olefins. Grubbs and Kroll suggested that this behavior was compatible with the idea that the reduction proceeded predominantly inside the polymer beads and that the solvent channels in the polymer restricted the entrance of the olefins to the catalytic site. However, as in the case with enzymes, the changed specificity of the rhodium catalyst could be attributed also to changed chemical properties caused by changes in coordination geometry, etc.

Other homogeneous catalysts and supports have been reported.[461-463]

## Other Immobilized Synthetic and Natural Compounds

Over the last twenty-five years, a vast number of compounds differing widely in size, complexity, and biological significance have been immobilized, mostly by covalent attachment to water-insoluble functionalized polymers. No attempt is made to even briefly summarize the different specific sub-

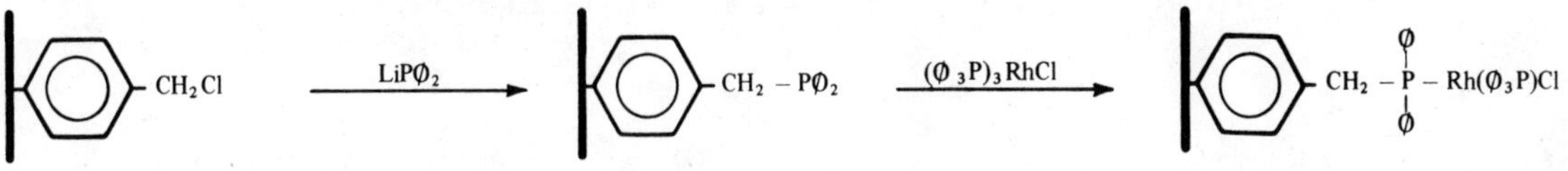

$$\text{P}-C_6H_4-CH_2Cl \xrightarrow{LiP\phi_2} \text{P}-C_6H_4-CH_2-P\phi_2 \xrightarrow{(\phi_3P)_3RhCl} \text{P}-C_6H_4-CH_2-P(\phi)_2-Rh(\phi_3P)Cl$$

$\phi = C_6H_5$

(82)

stances that have been immobilized, the reasons for their attachment, or the water-insoluble supports employed. Macromolecular substances such as antigens, antibodies, protease inhibitors,[464-466] hormones,[467-469] carbohydrate-binding proteins,[470] nucleic acids,[471,472] and glycosaminoglycans[473] have been frequently immobilized. Likewise, a large number of smaller-sized molecules such as amino acids,[474,475] vitamins,[476] enzyme inhibitors, and enzyme substrates have been covalently attached to water-insoluble matrices. Excellent reviews on immobilized antigens[32,114] and antibodies and on "affinity chromatography"[477,478] are available. It is interesting to note that although the covalent attachment of specific enzyme inhibitors and substrates to water-insoluble supports for the purpose of enzyme isolation was suggested and achieved in the early '50's,[479] it has only recently been "rediscovered" and developed by many biochemists.

Finally, this chapter on related topics cannot be closed without mentioning the work of Merrifield and others on protein synthesis[480] and degradation.[481,482] The concept and use of solid phase chemistry seems to be at the threshold of even greater scientific and technological developments.

# ABBREVIATIONS, SYMBOLS, AND TRADE NAMES

| | |
|---|---|
| A | Adenosine |
| ADP | Adenosine 5′-diphosphate |
| AE | Aminoethyl group |
| AMP | Adenosine 5′-phosphate |
| ATEE | *N*-Acetyl-L-tyrosine ethyl ester |
| ATP | Adenosine 5′-triphosphate |
| BAA | α-*N*-Benzoyl-L-arginine amide |
| BAEE | α-*N*-Benzoyl-L-arginine ethyl ester |
| CCP | Cytidine 2′,3′-cyclic phosphate |
| CDAPC | 1-Cyclohexyl-3-(3′-dimethylamino-propyl)carbodiimide |
| CM | Carboxymethyl group |
| CMC | 1-Cyclohexyl-3-(2-morpholinoethyl)-carbodiimide metho-*p*-toluene-sulfonate |
| DCC | *N,N*′-Dicyclohexylcarbodiimide |
| DEAE | Diethylaminoethyl group |
| L-DOPA | L-3,4-Dihydroxyphenylalanine |
| E | Enzyme (usually refers to concentra tion) |
| ECTEALO | Epichlorohydrin triethanolamine group |
| EDAPC | 1-Ethyl-3-(3′-dimethylaminopropyl) carbodiimide |
| Enzacryl® | Polyacrylamide-based water-insoluble matrices; trademark, Koch-Light Laboratories, Ltd. |
| FAD | Flavin adenine dinucleotide |
| FMN | Riboflavin 5′-phosphate |
| G | Guanosine |
| I | Inosine |
| IMP | Inosine 5′-phosphate |
| $K_m$ | Michaelis constant |
| $K'_m$ | Apparent Michaelis constant |
| k | Catalytic rate constant (decomposition of ES complex) |
| $NAD^+$ | Nicotinamide-adenine dinucleotide (oxidized form) |
| $NADP^+$ | Nicotinamide-adenine dinucleotide phosphate (oxidized form) |
| NMN | Nicotinamide mononucleotide |
| Poliodal | Homopolymer of 4-iodobutyl methacry-late or 2-iodoethyl methacrylate |
| S | Substrate (usually refers to concentra-tion) |
| Sephadex® | Spherical dextran gel particles; trade-mark Pharmacia Fine Chemicals, Inc. |
| Sepharose® | Spherical agarose gel particles; trade-mark Pharmacia Fine Chemicals, Inc. |
| Silastic® | Silicone rubber; trademark, Dow Corn-ing Corp. |
| S-MDA | Starch-methylenedianiline resin |
| Span® | Nonionic surfactants composed of partial esters of fatty acids (lauric, palmitic, stearic, and oleic) and hexitol anhy-drides (hexitans and hexides) derived from sorbitol; trademark, Atlas Chem-ical Industries, Inc. |
| TAME | α-*N*-*p*-Toluenesulfonyl-L-arginine methyl ester |
| TEMED | *N,N,N′,N′*-Tetramethylethylene-diamine |
| Tris | Tris(hydroxymethyl)aminomethane |
| Tween® | Nonionic surfactants derived from the Span surfactants by adding poly-oxyethylene chains to the nonesteri-fied hydroxyls; trademark, Atlas Chem-ical Industries, Inc. |
| U | Uridine |
| WRK | *N*-Ethyl-5-phenylisoxazolium-3′-sulfon-ate, Woodward's Reagent K |

## REFERENCES


1. **Katchalski, E., Silman, I., and Goldman, R.,** Effect of the microenvironment on the mode of action of immobilized enzymes, *Adv. Enzymol., Relat. Areas Mol. Biol.,* 34, 445, 1971.
2. **Goldman, R., Goldstein, L., and Katchalski, E.,** Water-insoluble enzyme derivatives and artificial enzyme membranes, in *Biochemical Aspects of Reactions on Solid Supports,* Stark, G. R., Ed., Academic Press, New York, 1971, 1.
3. **Melrose, G. J. H.,** Insolubilized enzymes; biochemical applications of synthetic polymers, *Rev. Pure Appl. Chem.,* 21, 83, 1971.
4. **Mosbach, K.,** Enzymes bound to artificial matrixes, *Sci. Am.,* 224, 26, 1971.
5. **Friedberg, F.,** Affinity chromatography and insoluble enzymes, *Chromatogr. Rev.,* 14, 121, 1971.
6. **Weetall, H. H.,** Enzymes immobilized on inorganic carriers, *Res./Develop.,* 22, 18, 1971.
7. **Falb, R. D. and Grode, G. A.,** Covalent binding of proteins to solid surfaces, *Fed. Proc.,* 30, 1688, 1971.
8. **Brown, E. and Racois, A.,** Insolubilized enzymes, *Bull. Soc. Chim. Fr.,* No. 12, 4613, 1971.
9. **Rogovin, Z. A. and Virnik, A. D.,** Biologically active cellulosic materials, in *Cellulose and Cellulose Derivatives,* IV, Bikales, N. M. and Segal, L., Eds., John Wiley & Sons, New York, 1971, 1333.
10. **Park, Y. K.,** Application of insoluble enzymes, *Biol. Inst. Technol. Aliment., Sao Paulo,* No. 26, 73, 1971 (*C. A.,* 76, 1080g, 1972).
11. **Martiny, S.,** Water-insoluble enzyme derivatives, *Dan. Kemi,* 52, 131, 1971 (*C. A.,* 76, 22252e, 1972).
12. **Pechan, Z.,** Preparation and use of insoluble enzymes, *Chem. Lide,* No. 2, 3, 1971 (*C. A.,* 75, 45122f, 1971).
13. **Gryszkiewicz, J.,** Insoluble enzymes, *Folia Biol. (Warsaw),* 19, 119, 1971 (*C. A.,* 75, 84388s, 1971).
14. **Goldstein, L.,** Water-insoluble derivatives of proteolytic enzymes, in *Methods Enzymol.,* Vol. 19, Perlman, G. E. and Lorand, L., Eds., Academic Press, New York, 1970, 935.
15. **Crook, E. M., Brocklehurst, K., and Wharton, C. W.,** Cellulose-insolubilized enzymes, in *Methods Enzymol.,* Vol. 19, Perlman, G. E. and Lorand, L., Eds., Academic Press, New York, 1970, 963.
16. **Alexander, B. and Engle, A. M.,** Water-insoluble thrombin, in *Methods Enzymol.,* Vol. 19, Perlman, G. E. and Lorand, L., Eds., Academic Press, New York, 1970, 978.
17. **Katchalski, E.,** Synthetic approach to the study of microenvironmental effects of enzyme action, in *Structure-Function Relationships of Proteolytic Enzymes,* Desnuelle, P., Neurath, H., and Ottesen, M., Eds., Academic Press, New York, 1970, 198.
18. **Emery, A. N. and Kent, C. A.,** Development of insolubilized enzymes, *Birmingham Univ. Chem. Eng.,* 21, 71, 1970.
19. **McLaren, A. D. and Packer, L.,** Some aspects of enzyme reactions in heterogeneous systems, *Adv. Enzymol., Relat. Areas Mol. Biol.,* 33, 245, 1970.
20. **Lindsey, A. S.,** Polymeric enzymes and enzyme analogs, in *Reviews in Macromolecular Chemistry,* Vol. 4, Butler, G. B. and O'Driscoll, K. F., Eds., Marcel Dekker, Inc., New York, 1970, 1.
21. **Chibata, I.,** Preparation and utilization of water-insoluble enzymes, *Yuki Gosei Kagaku Kyokai Shi,* 28, 471, 1970 (*C. A.,* 73, 52377e, 1970).
22. **Crook, E. M.,** Enzymes on solid matrixes, in *Metab. Regul. Enzyme Action, Fed. Eur. Biochem. Soc. Meet., 6th,* 1969, Sols, A., Ed., Academic Press, London, 1970, 297.
23. **Goldstein, L.,** Use of water-insoluble enzyme derivatives in synthesis and separation, in *Ferment. Advan., Pap. Int. Ferment. Symp., 3rd,* Perlman, D., Ed., Academic Press, New York, 1969, 391.
24. **Katchalski, E.,** Preparation and properties of enzymes immobilized in artificial membranes, in *Symmetry Funct. Biol. Syst. Macromol. Level, Proc. Nobel Symp., 11th,* Engstrom, A., Ed., Almqvist Wiksell, Stockholm, Sweden, 1969, 283 (*C. A.,* 72, 18761v, 1970).
25. **Chibata, I. and Tosa, T.,** Insoluble enzymes, *Kagaku To Seibutsu,* 7, 147, 1969 (*C. A.,* 72, 51113h, 1970).
26. **Goldstein, L. and Katchalski, E.,** Use of water-insoluble enzyme derivatives in biochemical analysis and separation, *Fresenius' Z. Anal. Chem.,* 243, 375, 1968.
27. **Kay, G.,** Insolubilized enzymes, *Process Biochem.,* 3, 36, 1968.
28. **Mark, H. F.,** Synthetic polymers in the medical sciences, *Pure Appl. Chem.,* 16, 201, 1968.
29. **Wiseman, A. and Gould, B.,** New enzymes for industry, *New Sci.,* 38, 66, 1968.

29a. **Crook, E. M.,** Enzymes attached to solid matrices, *Biochem. J.,* 107, 1p, 1968.

29b. **Manecke, G.,** Methods of attaching enzymes to various polymers and the effects on the properties of these enzymes, *Biochem. J.,* 107, 2p, 1968.

30. **Tosa, T. and Chibata, I.,** Properties and uses of water-insoluble enzymes, *Hakko Kyokaishi,* 25, 49, 1967 (*C. A.,* 67, 8076h, 1967).
31. **Chibata, I. and Tosa, T.,** Insoluble enzymes, *Kobunshi,* 16, 823, 1967 (*C. A.,* 69, 200g, 1968).
32. **Silman, I. H. and Katchalski, E.,** Water-insoluble derivatives of enzymes, antigens, and antibodies, *Ann. Rev. Biochem.,* 35, 873, 1966.
33. **Chibata, I. and Tosa, T.,** Preparation of insoluble enzymes, *Tampakushitsu Kakusan Koso,* 11, 23, 1966 (*C. A.,* 67, 79063n, 1967).

34. **Katchalski, E.**, Polyamino acids as protein models, *Proc. Plenary Sessions, 6th Internat. Congress of Biochem.*, I. U. B., 33, 81, 1964.
35. **Grubhofer, N. and Schleith, L.**, Modified ion-exchange resins as specific adsorbents, *Naturwissenschaften*, 40, 508, 1953.
36. **Grubhofer, N. and Schleith, L.**, Coupling of proteins on diazotized polyaminostyrene, *Hoppe-Seyler's Z. Physiol. Chem.*, 297, 108, 1954.
37. **Brandenberger, H.**, Methods for linking enzymes to insoluble carriers, *Angew. Chem.*, 67, 661, 1955.
38. **Brandenberger, H.**, Methods for linking enzymes to insoluble carriers, *Rev. Ferment. Ind. Aliment.*, 11, 237, 1956.
39. **Zingaro, R. A. and Uziel, M.**, Preparation and properties of active, insoluble alkaline phosphatase, *Biochim. Biophys. Acta*, 213, 371, 1970.
40. **Wilson, R. J. H., Kay, G., and Lilly, M. D.**, The preparation and properties of pyruvate kinase attached to porous sheets, and the operation of a two-enzyme continuous-feed reactor, *Biochem. J.*, 109, 137, 1968.
41. **Manecke, G.**, Serologically active protein resins and enzyme resins, *Naturwissenschaften*, 51, 25, 1964.
42. **Barker, S. A., Somers, P. J., and Epton, R.**, Preparation and stability of exo-amylolytic enzymes chemically coupled to microcrystalline cellulose, *Carbohyd. Res.*, 9, 257, 1969.
43. **Brown, E. and Racois, A.**, Insoluble enzymes. III. Preparation and properties of an insoluble derivative of chymotrypsinogen A and of chymotrypsins α and δ, *Bull. Soc. Chim. Fr.*, No. 12, 4357, 1971.
44. **Green, M. L. and Crutchfield, G.**, Studies on the preparation of water-insoluble derivatives of rennin and chymotrypsin and their use in the hydrolysis of casein and clotting of milk, *Biochem. J.*, 115, 183, 1969.
45. **Glassmeyer, C. K. and Ogle, J. D.**, Properties of an insoluble form of trypsin, *Biochemistry*, 10, 786, 1971.
45a. **Ogle, J. D. and Glassmeyer, C. K.**, Preparation and properties of insolubilized trypsin, *Abstr. 162nd Meet. Am. Chem. Soc., Div. Org. Coatings Plastic Chem.*, 31, 362, 1971.
46. **Riesel, E. and Katchalski, E.**, Preparation and properties of water-insoluble derivatives of urease, *J. Biol. Chem.*, 239, 1521, 1964.
47. **Goldstein, L., Pecht, M., Blumberg, S., Atlas, D., and Levin, Y.**, Water-insoluble enzymes. Synthesis of a new carrier and its utilization for preparation of insoluble derivatives of papain, trypsin, and subtilopeptidase A, *Biochemistry*, 9, 2322, 1970.
48. **Axen, R. and Ernback, S.**, Chemical fixation of enzymes to cyanogen halide activated polysaccharide carriers, *Eur. J. Biochem.*, 18, 351, 1971.
49. **Hornby, W. E., Lilly, M. D., and Crook, E. M.**, The preparation and properties of ficin chemically attached to carboxymethylcellulose, *Biochem. J.*, 98, 420, 1966.
50. **Brown, H. D., Chattopadhyay, S. K., and Patel, A.**, Sarcoplasmic reticulum ATPase on a solid support, *Biochem. Biophys. Res. Commun.*, 25, 304, 1966.
51. **Brown, H. D., Patel, A. B., Chattopadhyay, S. K., and Pennington, S. N.**, Cellulose-matrix supported biological substances, *Enzymologia*, 35, 215, 1968.
52. **Bar-Eli, A. and Katchalski, E.**, A water-insoluble trypsin derivative and its use as a trypsin column, *Nature*, 188, 856, 1960.
53. **Glazer, A. N., Bar-Eli, A., and Katchalski, E.**, Preparation and characterization of polytyrosyl trypsin, *J. Biol. Chem.*, 237, 1832, 1962.
53a. **Levin, Y., Pecht, M., Goldstein, L., and Katchalski, E.**, Polyanionic and polycationic derivatives of chymotrypsin, *7th Int. Cong. Biochem., Tokyo, Int. Union Biochem.*, 4, 800, Abstr. F-207, 1967.
54. **Engel, A. M. and Alexander, B.**, Studies of thrombin formation with an insoluble derivative of prothrombin, *J. Biol. Chem.*, 246, 1213, 1971.
55. **Bar-Eli, A. and Katchalski, E.**, Preparation and properties of water-insoluble derivatives of trypsin, *J. Biol. Chem.*, 238, 1690, 1963.
56. **Epton, R. and Thomas, T. H.**, Improving Nature's catalysts, *Aldrichimica Acta*, 4, 61, 1971.
57. **Mitz, M. A. and Summaria, L. J.**, Synthesis of biologically active cellulose derivatives of enzymes, *Nature*, 189, 576, 1961.
58. **Manecke, G. and Singer, S.**, Chemical transformations of polyaminostyrene, *Makromol. Chem.*, 36, 119, 1959.
59. **Weetall, H. H.**, Trypsin and papain covalently coupled to porous glass: preparation and characterization, *Science*, 166, 615, 1969.
60. **Weetall, H. H.**, Alkaline phosphatase insolubilized by covalent linkage to porous glass, *Nature*, 223, 959, 1969.
61. **Kay, G. and Crook, E. M.**, Coupling of enzymes to cellulose using chloro-*s*-triazines, *Nature*, 216, 514, 1967.
62. **Kay, G., Lilly, M. D., Sharp, A. K., and Wilson, R. J. H.**, Preparation and use of porous sheets with enzyme action, *Nature*, 217, 641, 1968.
63. **Wilson, R. J. H., Kay, G., and Lilly, M. D.**, The preparation and kinetics of lactate dehydrogenase attached to water-insoluble particles of sheets, *Biochem. J.*, 108, 845, 1968.
64. **Sharp, A. K., Kay, G., and Lilly, M. D.**, The kinetics of β-galactosidase attached to porous cellulose sheets, *Biotechnol. Bioeng.*, 11, 363, 1969.
65. **Wilson, R. J. H. and Lilly, M. D.**, Preparation and use of insolubilized amyloglucosidase for the production of sweet glucose liquors, *Biotechnol. Bioeng.*, 11, 349, 1969.

66. **Self, D. A., Kay, G., and Lilly, M. D.,** The conversions of benzyl penicillin to 6-aminopenicillanic acid using an insoluble derivative of penicillin amidase, *Biotechnol. Bioeng.,* 11, 337, 1969.
67. **Manecke, G. and Singer, S.,** Reactions of methacrylic acid fluoroanilide copolymers, *Makromol. Chem.,* 39, 13, 1960.
68. **Levin, Y., Pecht, M., Goldstein, L., and Katchalski, E.,** A water-insoluble polyanionic derivative of trypsin. I. Preparation and properties, *Biochemistry,* 3, 1905, 1964.
69. **Goldstein, L., Levin, Y., and Katchalski, E.,** A water-insoluble polyanionic derivative of trypsin. II. Effect of the polyelectrolyte carrier on the kinetic behavior of the bound trypsin, *Biochemistry,* 3, 1913, 1964.
70. **Fritz, H., Hochstrasser, K., Werle, E., Brey, B., and Gebhardt, M.,** Identification and preparative separation of proteinase inhibitors and of proteinases by water-insoluble enzyme and inhibitor resins, *Fresenius' Z. Anal. Chem.,* 243, 452, 1968.
71. **Axen, R., Porath, J., and Ernback, S.,** Chemical coupling of peptides and proteins to polysaccharides by means of cyanogen halides, *Nature,* 214, 1302, 1967.
72. **Porath, J., Axen, R., and Ernback, S.,** Chemical coupling of proteins to agarose, *Nature,* 215, 1491, 1967.
73. **Barker, S. A., Somers, P. J., Epton, R., and McLaren, J. V.,** Crosslinked polyacrylamide derivatives (Enzacryls) as water-insoluble carriers of amylolytic enzymes, *Carbohyd. Res.,* 14, 287, 1970.
74. **Barker, S. A., Somers, P. J., and Epton, R.,** Recovery and re-use of water-insoluble amylase derivatives, *Carbohyd. Res.,* 14, 323, 1970.
75. **Barker, S. A. and Epton, R.,** Water-insoluble enzymes, *Process Biochem.,* 5, 14, 1970.
76. **Epton, R., McLaren, J. V., and Thomas, T. H.,** Enzyme insolubilization with cross-linked polyacryloylamino-acetaldehyde dimethylacetal, *Biochem. J.,* 123, 21p, 1971.
77. **Inman, J. K. and Dintzis, H. M.,** The derivatization of cross-linked polyacrylamide beads. Controlled introduction of functional groups for the preparation of special-purpose, biochemical adsorbents, *Biochemistry,* 8, 4074, 1969.
78. **Weston, P. D. and Avrameas, S.,** Proteins coupled to polyacrylamide beads using glutaraldehyde, *Biochem. Biophys. Res. Commun.,* 45, 1574, 1971.
79. **Brown, E., Racois, A., and Gueniffey, H.,** Preparation and properties of urease derivatives insoluble in water, *Tetrahedron Lett.,* No. 25, 2139, 1970.
80. **Brown, E., Racois, A., Joyeau, R., Bonte, A., and Rioual, J.,** Preparation and properties of insoluble chymotrypsin derivatives, *C. R. Acad. Sci., Ser. C,* 273, 668, 1971.
81. **Brown, E., Racois, A., and Gueniffey, H.,** Insoluble enzymes. I. Preparation of new macromolecular supports: the polymethacrylates of $\omega$-iodo-*n*-alcohols (poliodals), *Bull. Soc. Chim. Fr.,* No. 12, 4341, 1971.
82. **Brown, E. and Racois, A.,** Insoluble enzymes. II. Preparation and properties of an insoluble derivative of urease and trypsin, *Bull. Soc. Chim. Fr.,* No. 12, 4351, 1971.
83. **Barker, S. A., Emery, A. N., and Novais, J. M.,** Enzyme reactors for industry, *Process Biochem.,* 5, 11, 1971.
84. **Hider, S. and Marchessault, R. H.,** Studies on alcohol-modified transition metal polymerization catalysts. II. Interaction of $TiCl_4$ with cellulose and model compounds, *J. Polym. Sci., Part C,* No. 11, 97, 1965.
85. **Axen, R., Vretblad, P., and Porath, J.,** The use of isocyanides for the attachment of biologically active substances to polymers, *Acta Chem. Scand.,* 25, 1129, 1971.
86. **Vretblad, P. and Axen, R.,** Covalent fixation of pepsin to agarose derivatives, *FEBS Lett.,* 18, 254, 1971.
87. **Ugi, I.,** The $\alpha$-addition of immonium ions and anions to isonitriles coupled with secondary reactions, *Angew. Chem.,* 74, 9, 1962.
88. **Stahmann, M. A. and Becker, R. R.,** A new method for adding amino acids and peptides to proteins, *J. Am. Chem. Soc.,* 74, 2695, 1952.
89. **Kirimura, J. and Yoshida, T.,** Water-insoluble copolymer of acylase with an amino acid anhydride, U.S. 3, 243, 356, March 29, 1966.
90. **Wykes, J. R., Dunnill, P., and Lilly, M. D.,** Immobilization of $\alpha$-amylase by attachment to soluble support materials, *Biochim. Biophys. Acta,* 250, 522, 1971.
91. **O'Neill, S. P., Wykes, J. R., Dunnill, P., and Lilly, M. D.,** An ultrafiltration-reactor system using a soluble immobilized enzyme, *Biotechnol. Bioeng.,* 13, 319, 1971.
92. **Conte, A. and Lehmann, K.,** Fixation of proteolytic enzymes on polymethacrylic acid anhydride, *Hoppe-Seyler's Z. Physiol. Chem.,* 352, 533, 1971.
93. **Brown, H. D., Patel, A. B., Chattopadhyay, S. K., and Pennington, S. N.,** Support matrixes for apyrase, *Enzymologia,* 35, 233, 1968.
94. **Patel, R. P., Lopiekes, D. V., Brown, S. P., and Price, S.,** Derivatives of proteins. II. Coupling of $\alpha$-chymotrypsin to carboxyl-containing polymers by use of N-ethyl-5-phenylisoxazolium-3′-sulfonate, *Biopolymers,* 5, 577, 1967.
95. **Axen, R., Myrin, P. A., and Janson, J. C.,** Chemical fixation of chymotrypsin to water-insoluble crosslinked dextran (Sephadex) and solubilization of the enzyme derivatives by means of dextranase, *Biopolymers,* 9, 401, 1970.
96. **Stasiw, R. O., Brown, H. D., and Hasselberger, F. X.,** Cholinesterase bonded to paper, *Can. J. Biochem.,* 48, 1314, 1970.
97. **Lilly, M. D.,** Stability of immobilized $\beta$-galactosidase on prolonged storage, *Biotechnol. Bioeng.,* 13, 589, 1971.
98. **Wheeler, K. D., Edwards, B. A., and Whittam, R.,** Some properties of two phosphates attached to insoluble cellulose matrices, *Biochim. Biophys. Acta,* 191, 187, 1969.

99. **Smith, N. L. and Lenhoff, H. M.,** Enzyme-cellophane membranes; covalently linked, *Fed. Proc.*, 30, Abstr. 1072, 1971.
100. **Balcom, J., Foulkes, P., Olson, N. F., and Richardson, T.,** Immobilized catalase, *Process Biochem.*, 6, 42, 1971.
101. **Hornby, W. E. and Filippusson, H.,** The preparation of trypsin chemically attached to nylon tubes, *Biochim. Biophys. Acta*, 220, 343, 1970.
102. **Filippusson, H. and Hornby, W. E.,** The preparation and properties of yeast β-fructofuranosidase chemically attached to polystyrene, *Biochem. J.*, 120, 215, 1970.
103. **Weetall, H. H.,** Insolubilized L-asparaginase implant: a preliminary report, *J. Biomed. Mater. Res.*, 4, 597, 1970.
104. **Weetall, H. H. and Hersh, L. S.,** Preparation and characterization of glucose oxidase covalently linked to nickel oxide, *Biochim. Biophys. Acta*, 206, 54, 1970.
105. **Julliard, J. H., Godinot, C., and Gautheron, D. C.,** Some modifications of the kinetic properties of bovine liver glutamate dehydrogenase (NAD(P)) covalently bound to a solid matrix of collagen, *FEBS Lett.*, 14, 185, 1971.
106. **Mosbach, K.,** Matrix-bound enzymes. Part I. The use of different acrylic copolymers as matrices, *Acta Chem. Scand.*, 24, 2084, 1970.
107. **Mosbach, K. and Mattiasson, B.,** Matrix-bound enzymes. Part II. Studies on a matrix-bound two-enzyme-system, *Acta Chem. Scand.*, 24, 2093, 1970.
108. **Marshall, D. L. and Falb, R. D.,** Preparation of insolubilized hexokinase, *Abstr. 162nd Meet. Am. Chem. Soc., Div. Org. Coatings Plastic Chem.*, 31, 363, 1971.
109. **Mattiasson, B. and Mosbach, K.,** Studies on a matrix-bound three-enzyme system, *Biochim. Biophys. Acta*, 235, 253, 1971.
110. **Manecke, G.,** Reactive polymers and their use for the preparation of antibody and enzyme resins, *Pure Appl. Chem.*, 4, 507, 1962.
111. **Manecke, G. and Günzel, G.,** Application of a nitrated copolymer from methacrylic acid and methacrylic acid-*m*-fluoroanilide to the preparation of enzyme resins, to the resolution of racemic compounds, and to tanning experiments, *Makromol. Chem.*, 51, 199, 1962.
112. **Monsan, P. and Durand, G.,** New preparation of enzymes fixed on inorganic supports, *C. R. Acad. Sci., Ser. C*, 273, 33, 1971.
113. **Weetall, H. H.,** Storage stability of water-insoluble enzymes covalently coupled to organic and inorganic carriers, *Biochim. Biophys. Acta*, 212, 1, 1970.
114. **Weliky, N. and Weetall, H. H.,** The chemistry and use of cellulose derivatives for the study of biological systems, *Immunochemistry*, 2, 293, 1965.
115. **Weibel, M. K. and Bright, H. J.,** Insolubilized enzymes, *Biochem. J.*, 124, 801, 1971.
116. Anonymous, Bonded enzymes found more active, stable, *Biochem. J.*, 124, 801, 1971.
117. **Lynn, J. and Falb, R. D.,** Active enzyme polymer adducts formed between aminoethylcellulose and aldolase, or glyceraldehyde-3-phosphate dehydrogenase or fructose-1,6-diphosphatase, *Abstr. 158th Meet. Am. Chem. Soc., Div. Biol. Sci.*, paper 298, 1969.
118. **Erlanger, B. G., Isambert, M. F., and Michelson, A. M.,** Insoluble bacterial luciferases: a new approach to some problems in bioluminescence, *Biochem. Biophys. Res. Commun.*, 40, 70, 1970.
119. **Mezzasoma, I. and Turano, C.,** Binding of enzymes to insoluble supports by the carbodiimide method. Evaluation of the method, *Boll Soc. Ital. Biol. Sper.*, 47, 407, 1971.
120. **Weetall, H. H. and Baum, G.,** Continuous processing of D,L amino acids with an insolubilized L-amino acid oxidase, *Abstr. 158th Meet. Am. Chem. Soc., Div. Biol. Chem.*, paper 153, 1969.
121. **Weetall, H. H. and Baum, G.,** Preparation and characterization of insolubilized L-amino acid oxidase, *Biotechnol. Bioeng.*, 12, 399, 1970.
122. **Wykes, J. R., Dunnill, P., and Lilly, M. D.,** Conversion of tyrosine to L-dihydroxyphenylalanine using immobilized tyrosinase, *Nature, New Biol.*, 230, 187, 1971.
123. **Surinov, B. P. and Manoilov, S. E.,** Production and properties of insoluble compounds of certain enzymes with cellulose, *Biokhimiya*, 31, 387, 1966 (Engl. trans., 31, 387, 1966).
124. **Yaglom, D. L., Virnik, A. D., and Rogovin, Z. A.,** Synthesis of graft copolymers of cellulose and catalase, *Vysokomol. Soedin., Ser. B*, 11, 287, 1969 (*C. A.*, 71, 40442t, 1969).
125. **Weliky, N., Brown, F. S., and Dale, E. C.,** Carrier-bound proteins: prcperties of peroxidase bound to insoluble carboxymethylcellulose particles, *Arch. Biochem. Biophys.*, 131, 1, 1969.
126. **Weetall, H. H. and Weliky, N.,** A new technique for the enzymic detection of hydrogen peroxide, *Anal. Biochem.*, 14, 160, 1966.
127. **Bohnensack, R., Augustin, W., and Hofmann, E.,** Chemical coupling of hexokinase from yeast on Sephadex, *Experientia*, 25, 348, 1969.
128. **Hornby, W. E., Lilly, M. D., and Crook, E. M.,** Some changes in the reactivity of enzymes resulting from their chemical attachment to water-insoluble derivatives of cellulose, *Biochem. J.*, 107, 669, 1968.
129. **Hoffman, C. H., Harris, E., Chodroff, S., Michelson, S., Rothrock, J. W., Peterson, E., and Reuter, W.,** Polynucleotide phosphorylase covalently bound to cellulose and its use in the preparation of homopolynucleotides, *Biochem. Biophys. Res. Commun.*, 41, 710, 1970.

130. **Monsan, P. and Durand, G.**, Preparation of insolubilized invertase by adsorption on bentonite, *FEBS Lett.*, 16, 39, 1971.
131. **Lilly, M., Money, C., Hornby, W., and Crook, E. M.**, Enzymes on solid supports, *Biochem. J.*, 95, 45p, 1965.
132. **Epstein, C. J. and Anfinsen, C. B.**, The reversible reduction of disulfide bonds in trypsin and ribonuclease coupled to carboxymethylcellulose, *J. Biol. Chem.*, 237, 2175, 1962.
133. **Givol, D., Weinstein, Y., Gorecki, M., and Wilchek, M.**, A general method for the isolation of labelled peptides from affinity-labelled proteins, *Biochem. Biophys. Res. Commun.*, 38, 825, 1970.
134. **Axen, R., Carlsson, J., Janson, J.-C., and Porath, J.**, Ribonuclease chemically attached to beads of epichlorohydrin cross-linked agarose, *Enzymologia,* 41, 359, 1971.
135. **Litvak, S., Tarrago-Litvak, L., Carre, D. S., and Chapeville, F.**, The synthesis and properties of the Sepharose-bound tRNA nucleotidyltransferase, *Eur. J. Biochem.*, 24, 249, 1971.
136. **Kuriyama, Y. and Egami, F.**, Preparation and properties of water-insoluble ribonuclease $T_1$, *Seikagaku,* 38, 735, 1966 (*C. A.*, 66, 72694g, 1967).
137. **Lee, J. C.**, Preparation and properties of water-insoluble derivatives of ribonuclease $T_1$, *Biochim. Biophys. Acta,* 235, 435, 1971.
138. **Baum, G., Ward, F. B., and Weetall, H. H.**, Acetylcholinesterase-siloxane-glass adducts, *Abstr. 162nd Meet. Am. Chem. Soc., Div. Org. Coatings Plastics Chem.*, 31, 352, 1971.
139. **Axen, R., Heilbronn, E., and Winter, A.**, Preparation and properties of cholinesterase covalently bound to Sepharose, *Biochim. Biophys. Acta,* 191, 478, 1969.
140. **Patel, A. N., Pennington, S. N., and Brown, H. D.**, Insoluble matrix-supported apyrase, deoxyribonuclease and cholinesterase, *Biochim. Biophys. Acta,* 178, 626, 1969.
141. **Grove, M. J., Strandberg, G. W., and Smiley, K. L.**, Steroid esterase bound to porous glass beads, *Biotechnol. Bioeng.*, 13, 709, 1971.
142. **Neurath, R. A. and Weetall, H. H.**, Deoxyribonuclease I covalently coupled to porous glass, *FEBS Lett.*, 8, 253, 1970.
143. **Omenn, G. S., Ontjes, D. A., and Anfinsen, C. B.**, Fractionation of antibodies against staphylococcal nuclease on Sepharose immunoadsorbents, *Nature,* 225, 189, 1970.
144. **Andria, G., Taniuchi, H., and Cone, J. L.**, The specific binding of three fragments of staphylococcal nuclease, *J. Biol. Chem.*, 246, 7421, 1971.
145. **Weetall, H. H.**, Preparation and characterization of an arylsulphatase insolubilized on porous glass, *Nature,* 232, 473, 1971.
146. **Ledingham, W. M. and Hornby, W. E.**, The action pattern of water-insoluble α-amylases, *FEBS Lett.*, 5, 118, 1969.
146a. **Maeda, H., Miyado, S., and Suzuki, H.**, Water-insoluble enzymes. 2. Continuous saccharification of liquified starch, *Hakko Kyokaishi,* 28, 391, 1970 (*C. A.*, 75, 62167s, 1971).
147. **Barker, S. A., Somers, P. J., and Epton, R.**, Preparation and properties of α-amylase chemically coupled to microcrystalline cellulose, *Carbohyd. Res.*, 8, 491, 1968.
148. **Manecke, G. and Forster, H. J.**, Reactive polystyrene-based polymers as carriers for proteins and enzymes, *Makromol. Chem.*, 91, 136, 1966.
149. **Axen, R. and Porath, J.**, Chemical coupling of enzymes to cross-linked dextran (Sephadex), *Nature,* 210, 367, 1966.
150. **Maeda, H. and Suzuki, H.**, Water-insoluble enzyme. I. General properties of CM-cellulose glucoamylase, *Nippon Nogei Kagaku Kaishi,* 44, 547, 1970 (*C. A.*, 74, 135442u, 1971).
151. **O'Neill, S. P., Dunnill, P., and Lilly, M. D.**, A comparative study of immobilized amyloglucosidase in a packed bed reactor and a continuous feed stirred tank reactor, *Biotechnol. Bioeng.*, 13, 337, 1971.
152. **Barker, S. A., Doss, S. H., Gray, C. J., Kennedy, J. F., Stacey, M., and Yeo, T. H.**, β-D-Glucosidase chemically bound to microcrystalline cellulose, *Carbohyd. Res.*, 20, 1, 1971.
153. **Robinson, P. J., Dunnill, P., and Lilly, M. D.**, Porous glass as a solid support for immobilization or affinity chromatography of enzymes, *Biochim. Biophys. Acta,* 242, 659, 1971.
154. **Woychik, J. H. and Wondolowski, V.**,Properties of β-D-galactosidase bound to glass, *Abstr. 162nd Meet. Am. Chem. Soc., Div. Agr. Food Chem.*, paper 14, 1971.
155. **Stimson, W. H. and Serafini-Fracassini, A.**, Hyaluronidase chemically bound to agarose, *FEBS Lett.*, 17, 318, 1971.
156. **Goldstein, L., Lifshitz, A., and Sokolovsky, M.**, Water-insoluble derivatives of naringinase, *Int. J. Biochem.*, 2, 446, 1971.
157. **Koelsch, R., Lasch, J., and Hanson, H.**, Covalent and noncovalent binding of leucine aminopeptidase to water-insoluble carriers, *Acta Biol. Med. Ger.*, 24, 833, 1970.
158. **Bennett, H. P. J., Elliott, D. F., Lowry, P. J., and McMartin, C.**, The complete hydrolysis of peptides by a mixture of Sepharose-bound peptidases, *Biochem. J.*, 125, 80p, 1971.
159. **Seki, T., Yang, H. Y. T., Levin, Y., Jenssen, T. A., and Erdős, E. G.**, Application of water-insoluble complexes of kininogenases, inhibitors and kinases to kinin research, *Adv. Exp. Med. Biol.*, 8, 23, 1970.
160. **Chiang, T. S., Erdős, E. G., Miwa, I., Tague, L. L., and Coalson, J. J.**, Isolation from a salivary gland of granules containing renin and kallikrein, *Circ. Res.*, 23, 507, 1968.
161. **Line, W. F., Kwong, A., and Weetall, H. H.**, Pepsin insolubilized by covalent attachment to glass: preparation and characterization, *Biochim. Biophys. Acta,* 242, 194, 1971.

162. **Ferrier, L. K., Richardson, T., and Olson, N. F.,** Continuous coagulation of skim milk with insoluble pepsin, *J. Dairy Sci.*, 54, 762, 1971.
163. **Fritz, H., Gebhardt, M., Meister, R., and Schult, H.,** Preparation of modified protease inhibitors using water-insoluble trypsin resins, *Hoppe-Seyler's Z. Physiol. Chem.*, 351, 1119, 1970.
164. **Ohno, Y. and Stahmann, M. A.,** Polyacrylamide derivatives of amino acid acylase and trypsin, *Macromolecules,* 4, 350, 1971.
165. **Wagner, T., Hsu, C.-J., and Kelleher, G.,** A new method for the attachment and support of active enzymes, *Biochem. J.*, 108, 892, 1968.
166. **Habeeb, A. F. S. A.,** Preparation of enzymically active, water-insoluble derivatives of trypsin, *Arch. Biochem. Biophys.*, 119, 264, 1967.
167. **Fritz, H., Gebhardt, M., Meister, R., Illchmann, K., and Hochstrasser, K.,** Isolation of protease inhibitors by water-insoluble trypsin-cellulose resins, *Hoppe-Seyler's Z. Physiol. Chem.*, 351, 571, 1970.
168. **Craven, G. R. and Gupta, V.,** Three-dimensional organization of the 30S ribosomal proteins from *Escherichia coli.* I. Preliminary classification of the proteins, *Proc. Natl. Acad. Sci. USA,* 67, 1329, 1970.
169. **Alexander, B., Rimon, A., and Katchalski, E.,** Effects of water-insoluble trypsin derivatives on fibrinogen, *Fed. Proc.*, 24, 804, 1965.
170. **Engel, A. and Alexander, B.,** Zymogen activation by insoluble trypsin and thrombin, *Fed. Proc.*, 24, 512, 1965.
171. **Lowey, S., Goldstein, L., and Luck, S.,** Isolation and characterization of a helical subunit from heavy meromyosin, *Biochem. Z.*, 345, 248, 1966.
172. **Rimon, A., Alexander, B., and Katchalski, E.,** Action of water-insoluble trypsin derivatives on prothrombin and related clotting factors, *Biochemistry,* 5, 792, 1966.
173. **Ong, E. B., Tsang, Y., and Perlmann, G. E.,** Action of water-insoluble trypsin derivatives on pepsinogen, *J. Biol. Chem.*, 241, 5661, 1966.
174. **Fritz, H., Schult, H., Neudecker, M., and Werle, E.,** Isolation of protease inhibitors, *Angew. Chem.*, 78, 775, 1966 (*Angew. Chem. Int. Ed. Engl.*, 5, 735, 1966).
175. **Goldstein, L., Lowey, S., Cohen, C., and Luck, S.,** Studies on the isolation and characterization of a helical subunit from heavy meromyosin, *Is. J. Chem.*, 5, 91p, 1967.
176. **Lowey, S., Goldstein, L., Cohen, C., and Luck, S. M.,** Proteolytic degradation of myosin and the meromyosins by a water-insoluble polyanionic derivative of trypsin, *J. Mol. Biol.*, 23, 287, 1967.
177. **Fritz, H., Trautschold, I., Haendle, H., and Werle, E.,** Chemistry and biochemistry of proteinase inhibitors from mammalian tissues, *Ann. NY Acad. Sci.*, 146, 400, 1968.
178. **Fritz, H., Brey, B., Schmal, A., and Werle, E.,** The use of water-insoluble derivatives of the trypsin-kallikrein inhibitor for the isolation of kallikrein and plasmin, *Hoppe-Seyler's Z. Physiol. Chem.*, 350, 617, 1969.
179. **Fritz, H., Gebhardt, M., Fink, E., Schramm, W., and Werle, E.,** The use of water-insoluble enzymes resins with polyanionic and polyamphoteric resin matrices for the isolation of protease inhibitors, *Hoppe-Seyler's Z. Physiol. Chem.*, 350, 129, 1969.
180. **Fritz, H., Schult, H., Hutzel, M., Wiedermann, M., and Werle, E.,** Protease inhibitors. IV. Isolation of protease inhibitors with water-insoluble enzyme-resin complexes, *Hoppe-Seyler's Z. Physiol. Chem.*, 348, 308, 1967.
181. **Fritz, H., Hutzel, M., and Werle, E.,** Protease inhibitors. VI. Identity of a protease inhibitor from bovine liver with the trypsin-kallikrein inhibitor, trasylol, *Hoppe-Seyler's Z. Physiol. Chem.*, 348, 950, 1967.
182. **Brown, E. and Racois, A.,** Preparation and properties of trypsin derivatives insoluble in water, *Tetrahedron Lett.*, No. 15, 1047, 1971.
183. **Gabel, D., Vretblad, P., Axen, R., and Porath, J.,** Insolubilized trypsin with activity in 8 M urea, *Biochim. Biophys. Acta,* 214, 561, 1970.
184. **Gabel, D., Steinberg, I. Z., and Katchalski, E.,** Changes in conformation of insolubilized trypsin and chymotrypsin, followed by fluorescence, *Biochemistry,* 10, 4661, 1971.
185. **Seki, T., Jenssen, T. A., Levin, Y., and Erdös, E. G.,** Active water-insoluble derivative of renin, *Nature,* 225, 864, 1970.
186. **Engel, A., Alexander, B., and Pechet, L.,** Preparation, properties, and activation of insoluble prothrombin and trypsinogen, *Fed. Proc.*, 25, 318, 1966.
187. Anonymous, Chromatography on bound enzyme precursors – zymogens, *Chem. Eng. News,* 49(40), 34, 1971.
188. **Kay, G. and Lilly, M. D.,** The chemical attachment of chymotrypsin to water-insoluble polymers using 2-amino-4,6-dichloro-*s*-triazine, *Biochim. Biophys. Acta,* 198, 276, 1970.
189. **Lilly, M. D. and Sharp, A. K.,** The kinetics of enzymes attached to water-insoluble polymers, *Chem. Eng.*, No 215, CE12, 1968.
190. **Surovtsev, V. I., Kozlov, L. V., and Antonov, V. K.,** Properties of chymotrypsin covalently bound with carboxymethylcellulose, *Dokl. Akad. Nauk SSSR,* 195, 1463, 1970 (*C. A.*, 74, 72067c, 1971).
191. **Surovtsev, V. I., Kozlov, L. V., and Antonov, V. K.,** Investigation of denaturation of $\alpha$-chymotrypsin linked by covalent bonds with carboxymethylcellulose, *Biokhimiya,* 36, 199, 1971 (Engl. trans., 36, 167, 1971).
192. **Feinstein, G.,** Isolation of chicken ovoinhibitor by affinity chromatography on chymotrypsin-Sepharose, *Biochim. Biophys. Acta,* 236, 73, 1971.

193. **Kasche, V., Lundgvist, H., Bergman, R., and Axen, R.,** A theoretical model describing steady-state catalysis by enzymes immobilized in spherical gel particles. Experimental study of α-chymotrypsin-Sepharose, *Biochem. Biophys. Res. Commun.,* 45, 615, 1971.
194. **Feinstein, G.,** Isolation of chymotrypsin inhibitors by affinity chromatography through chymotrypsin-Sepharose, in *Proceedings of the First International Conference on Proteinase Inhibitors,* 1970, Fritz, H., Ed., deGruyter, Berlin, 1971, 38 (*C. A.,* 75, 137083x, 1971).
195. **Cebra, J. J., Givol, D., Silman, H. I., and Katchalski, E.,** A two-stage cleavage of rabbit γ-globulin by a water-insoluble papain preparation followed by cysteine, *J. Biol. Chem.,* 236, 1720, 1961.
196. **Cebra, J. J., Givol, D., and Katchalski, E.,** Soluble complexes of antigen and antibody fragments, *J. Biol. Chem.,* 237, 751, 1962.
197. **Givol, D. and Sela, M.,** A comparison of fragments of rabbit antibodies and normal γ-globulin by the peptide-map technique, *Biochemistry,* 3, 451, 1964.
198. **Cebra, J. J.,** The effect of sodium dodecylsulfate on intact and insoluble papain hydrolyzed immune globulin, *J. Immunol.,* 92, 977, 1964.
199. **Jaquet, H. and Cebra, J. J.,** A comparison of two precipitating derivatives of rabbit antibody: fragment I dimer and the product of pepsin digestion, *Biochemistry,* 4, 954, 1965.
200. **Silman, I. H., Albu-Weissenberg, M., and Katchalski, E.,** Some water-insoluble papain derivatives, *Biopolymers,* 4, 441, 1966.
201. **Slayter, H. S. and Lowey, S.,** Substructure of the myosin molecule as visualized by electron microscopy, *Proc. Natl. Acad. Sci. USA,* 58, 1611, 1967.
202. **Kominz, D. R., Mitchell, E. R., Nihei, I., and Kay, C. M.,** The papain digestion of skeletal myosin A, *Biochemistry,* 4, 2373, 1965.
203. **Manecke, G. and Günzel, G.,** Polymeric isothiocyanates for the preparation of highly active resins for enzyme studies, *Naturwissenschaften,* 54, 531, 1967.
204. **Manecke, G., Günzel, G., and Forster, H. J.,** Enzymes covalently bound to various polymers and the effects on the properties of these enzymes, *J. Polym. Sci.,* Part C, No. 30, 607, 1970.
205. **Lilly, M. D., Hornby, W. E., and Crook, E. M.,** The kinetics of carboxymethylcellulose-ficin in packed beds, *Biochem. J.,* 100, 718, 1966.
206. **Hussain, Q. Z. and Newcomb, T. F.,** Preparation of water-insoluble thrombin, *Proc. Soc. Exp. Biol. Med.,* 115, 301, 1963.
207. **Owen, W. G. and Wagner, R. H.,** Preparation and properties of water-insoluble thrombin, *Am. J. Physiol.,* 220, 1941, 1971.
208. **Newcomb, T. F. and Hoshida, M.,** Factor V and thrombin, *Scand. J. Clin. Lab. Invest.,* 17, Suppl. 84, 61, 1965.
209. **Cohen, C., Slayter, H., Goldstein, L., Kucera, J., and Hall, C.,** Polymorphism in fibrinogen aggregates, *J. Mol. Biol.,* 22, 385, 1966.
210. **Rimon, S., Stupp, Y., and Rimon, A.,** Studies on the activation of plasminogen. III. Insoluble derivatives of the proactivator and the activator, *Can. J. Biochem.,* 44, 415, 1966.
211. **Wharton, C. W., Crook, E. M., and Brocklehurst, K.,** The preparation and some properties of bromelain covalently attached to *O*-(carboxymethyl)-cellulose, *Eur. J. Biochem.,* 6, 565, 1968.
212. **Wharton, C. W., Crook, E. M., and Brocklehurst, K.,** The nature of the perturbation of the Michaelis constant of the bromelain-catalysed hydrolysis of α-N-benzoyl-L-arginine ethyl ester consequent upon attachment of bromelain to *O*-(carboxymethyl)-cellulose, *Eur. J. Biochem.,* 6, 572, 1968.
213. **Cresswell, P. and Sanderson, A. R.,** Preparation, properties and substrate-exclusion effects of an insoluble pronase derivative, *Biochem. J.,* 119, 447, 1970.
214. **Cresswell, P. and Sanderson, A. R.,** Insoluble pronase, *Biochem. J.,* 117, 43p, 1970.
215. **Shaltiel, S., Mizrahi, R., Stupp, Y., and Sela, M.,** Reduction in immunological manifestations of 6-aminopenicillanic acid by treatment with water-insoluble pronase, *Eur. J. Biochem.,* 14, 509, 1970.
216. **Royer, G. P. and Green, G. M.,** Immobilized pronase, *Biochem. Biophys. Res. Commun.,* 44, 426, 1971.
217. **Royer, G. P. and Green, G. M.,** Immobilized pronase, *Abstr. 162nd Meet. Am. Chem. Soc., Div. Org. Coatings Plastics Chem.,* 31, 357, 1971.
218. **Rimon, A., Gutman, M., and Rimon, S.,** Studies on the activation of plasminogen. I. Preparation and properties of an insoluble derivative of streptokinase, *Biochim. Biophys. Acta,* 73, 301, 1963.
219. **Caviezel, O.,** Inactivation of streptokinase by polyaminostyrene, *Schweiz. Med. Wochenschr.,* 94, 1194, 1964 (*C. A.,* 62, 8068f, 1965).
220. **Steinbuch, M. and Prejaudier, L.,** Activation of plasminogen by a streptokinase coupled to a cellulose support, *Bibl. Haematol.,* No. 19, 169, 1964 (*C. A.,* 61, 9842c, 1964).
221. **Gabel, D. and v. Hofsten, B.,** Some properties of a bacterial proteinase chemically fixed to agarose, *Eur. J. Biochem.,* 15, 410, 1970.
222. **Hasselberger, F. X., Brown, H. D., Chattopadhyay, S. K., Mather, A. D., Stasiw, R. O., Patel, A. B., and Pennington, S. N.,** The preparation of insoluble, matrix-supported derivatives of asparaginase for use in cancer therapy, *Cancer Res.,* 30, 2736, 1970.

223. **Weetall, H. H. and Hersh, L. S.,** Urease covalently coupled to porous glass, *Biochim. Biophys. Acta,* 185, 464, 1969.
224. **Sundaram, P. V. and Hornby, W. E.,** Preparation and properties of urease chemically attached to nylon tube, *FEBS Lett.,* 10, 325, 1970.
225. **Sato, T., Mori, T., Tosa, T., and Chibata, I.,** Studies on immobilized enzymes. IX. Preparation and properties of aminoacylase covalently attached to halogenoacetylcelluloses, *Arch. Biochem. Biophys.,* 147, 788, 1971.
226. **Brown, H. D., Chattopadhyay, S. K., and Patel, A.,** Characteristics of an ATPase in membrane particles, "solubilized," and linked to a cellulose matrix, *Enzymologia,* 32, 205, 1967.
227. **Brown, H. D., Chattopadhyay, S. K., and Swaraj, K.,** Properties of solubilized and matrix-supported erythrocyte membrane ATPase, *Tex. J. Sci.,* 23, 235, 1971.
228. **Brown, H. D., Patel, A. B., and Chattopadhyay, S. K.,** Ethylene-maleic acid copolymer, aminopolystyrene and polyvinylamine matrices for potato apyrase, *Am. J. Bot.,* 55, 729, 1968.
229. **Brown, H. D., Patel, A. B., and Chattopadhyay, S. K.,** The properties of apyrase upon a solid support, *Plant Physiol.,* 41 (Suppl.), IXVI, 1966.
230. **Whittam, R., Edwards, B. A., and Wheeler, K. P.,** An approach to the study of enzyme action in artificial membranes, *Biochem. J.,* 107, 3p, 1968.
231. **Graubaum, H., Jeschkeit, H., and Schellenberger, A.,** Covalent binding of yeast pyruvate decarboxylase (EC 4.1.1.1) to aminoethylpolystyrene resin, *Z. Chem.,* 11, 107, 1971 (*C. A.,* 75, 30617x, 1971).
232. **Chan, W. W.-C.,** Matrix-bound protein subunits, *Biochem. Biophys. Res. Commun.,* 41, 1198, 1970.
233. **Denburg, J. and DeLuca, M.,** Purification of a specific tRNA by Sepharose-bound enzyme, *Proc. Natl. Acad. Sci. USA,* 67, 1057, 1970.
234. **Axen, R. and Vretblad, P.,** Binding of proteins to polysaccharides by means of cyanogen halides. Studies on cyanogen bromide treated Sephadex, *Acta Chem. Scand.,* 25, 2711, 1971.
235. **Wold, F.,** Bifunctional reagents, in *Methods Enzymol.,* Vol. 11, Hirs, C. H. W., Ed., Academic Press, New York, 1967, 617.
236. **Fasold, H., Klappenberger, J., Meyer, C., and Remold, H.,** Bifunctional reagents for the crosslinking of proteins, *Angew. Chem. Int. Ed., Engl.,* 10, 795, 1971.
237. **Jansen, E. F. and Olson, A. C.,** Properties and enzymatic activities of papain insolubilized with glutaraldehyde, *Arch. Biochem. Biophys.,* 129, 221, 1969.
238. **Jansen, E. F., Tomimatsu, Y., and Olson, A. C.,** Cross-linking of α-chymotrypsin and other proteins by reaction with glutaraldehyde, *Arch. Biochem. Biophys.,* 144, 394, 1971.
239. **Tomimatsu, Y., Jansen, E. F., Gaffield, W., and Olson, A. C.,** Physical chemical observations on the α-chymotrypsin glutaraldehyde system during formation of an insoluble derivative, *J. Colloid Interface Sci.,* 36, 51, 1971.
240. **Ottesen, M. and Svensson, B.,** Modification of papain by treatment with glutaraldehyde under reducing and non-reducing conditions, *C. R. Trav. Lab. Carlsberg,* 38, 171, 1971.
241. **Quiocho, F. A. and Richards, F. M.,** Intermolecular cross linking of a protein in the crystalline state: carboxypeptidase-A, *Proc. Natl. Acad. Sci. USA,* 52, 833, 1964.
242. **Quiocho, F. A. and Richards, F. M.,** The enzymic behavior of carboxypeptidase-A, in the solid state, *Biochemistry,* 5, 4062, 1966.
243. **Bishop, W. H., Quiocho, F. A., and Richards, F. M.,** The removal and exchange of metal ions in cross-linked crystals of carboxypeptidase-A, *Biochemistry,* 5, 4077, 1966.
244. **Marfey, P. S. and King, M. V.,** Chemical modification of ribonuclease A crystals. I. Reaction with 1,5-difluoro-2,4-dinitrobenzene, *Biochim. Biophys. Acta,* 105, 178, 1965.
245. **Haas, D. J.,** Preliminary studies on the denaturation of cross-linked lysozyme crystals, *Biophys. J.,* 8, 549, 1968.
246. **Chang, T. M. S.,** Stabilization of enzymes by microencapsulation with a concentrated protein solution or by microencapsulation followed by cross-linking with glutaraldehyde, *Biochem. Biophys. Res. Commun.,* 44, 1531, 1971.
247. **Avıameas, S. and Ternynck, T.,** The cross-linking of proteins with glutaraldehyde and its use for the preparation of immunoadsorbents, *Immunochemistry,* 6, 53, 1969.
248. **Avrameas, S.,** Coupling of enzymes to proteins with glutaraldehyde. Use of the conjugates for the detection of antigens and antibodies, *Immunochemistry,* 6, 43, 1969.
249. **Patramani, I., Katsiri, K., Pistevou, E., Kalogerakos, T., Pavlatos, M., and Evangelopoulos, A. E.,** Glutamic-aspartic transaminase – antitransaminase interaction: a method for antienzyme purification, *Eur. J. Biochem.,* 11, 28, 1969.
250. **Selegny, E., Broun, G., and Thomas, D.,** Enzymatically active model-membranes: experimental illustrations and calculations on the basis of diffusion-reaction kinetics of their functioning, of regulatory effects, of facilitated, retarded and active transports, *Physiol. Veg.,* 9, 25, 1971.
251. **Ogata, K., Ottesen, M., and Svendsen, I.,** Preparation of water-insoluble, enzymatically active derivatives of subtilisin type Novo by crosslinking with glutaraldehyde, *Biochim. Biophys. Acta,* 159, 403, 1968.
252. **Haynes, R. and Walsh, K. A.,** Enzyme envelopes on colloidal particles, *Biochem. Biophys. Res. Commun.,* 36, 235, 1969.

253. **Walsh, K. A., Houston, L. L., and Kenner, R. A.,** Chemical modification of bovine trypsinogen and trypsin, in *Structure-Function Relationships of Proteolytic Enzymes,* Desnuelle, P., Neurath, H., and Ottesen, M., Eds., Academic Press, New York, 1970, 56.
254. **Walsh, K. A., Pangburn, M., and Haynes, R.,** Enzyme monolayers attached to colloidal silica, *Abstr. 162nd Meet. Am. Chem. Soc., Div. Org. Coatings Plastics Chem.,* 31, 361, 1971.
255. **Gardner, D. L., Falb, R. D., Kim, B. C., and Emmerling, D. C.,** Possible uremic detoxification via oral-ingested microcapsules, *Trans. Am. Soc. Artif. Intern. Organs,* 17, 239, 1971.
256. **Gardner, D. L. and Emmerling, D.,** Microencapsulated stabilized urease, *Abstr. 162nd Meet. Am. Chem. Soc., Div. Organic Coatings Plastics Chem.,* 31, 366, 1971.
257. **Broun, G., Selegny, E., Minh, C. T., and Thomas, D.,** Facilitated transport of $CO_2$ across a membrane bearing carbonic anhydrase, *FEBS Lett.,* 7, 223, 1970.
258. **Goldman, R., Silman, H. I., Caplan, S. R., Kedem, O., and Katchalski, E.,** Papain membrane on a collodion matrix: preparation and enzymic behavior, *Science,* 150, 758, 1965.
259. **Goldman, R., Kedem, O., Silman, I. H., Caplan, S. R., and Katchalski, E.,** Papain-collodion membranes. I. Preparation and properties, *Biochemistry,* 7, 486, 1968.
260. **Goldman, R., Kedem, O., and Katchalski, E.,** Papain-collodion membranes. II. Analysis of the kinetic behavior of enzymes immobilized in artificial membranes, *Biochemistry,* 7, 4518, 1968.
261. **Goldman, R., Kedem, O., and Katchalski, E.,** Kinetic behavior of alkaline phosphatase-collodion membranes, *Biochemistry,* 10, 165, 1971.
262. **Selegny, E., Avrameas, S., Broun, G., and Thomas, D.,** Membranes with enzymic activity, synthesis of membranes with covalently bound enzymes; characterization of the catalytic activity by diffusion-reaction, *C. R. Acad. Sci., Ser. C,* 266, 1431, 1968.
263. **Broun, G., Selegny, E., Avrameas, S., and Thomas, D.,** Enzymatically active membranes: some properties of cellophane membranes supporting cross-linked enzymes, *Biochim. Biophys. Acta,* 185, 258, 1969.
264. **Selegny, E., Broun, G., and Thomas, D.,** Enzymes in an insoluble phase. Variation of enzymic activity as a function of substrate composition. Regulation of enzymic membranes, *C. R. Acad. Sci., Ser. D,* 269, 1330, 1969.
265. **Selegny, E., Broun, G., Geffroy, J., and Thomas, D.,** Determining the real Michaelis constant of an enzyme from the steady state of an enzyme-active membrane, *J. Chim. Phys. Physicochim. Biol.,* 66, 391, 1969.
266. **Avrameas, S. and Ternynck, T.,** Biologically active water-insoluble protein polymers. I. Their use for isolation of antigens and antibodies, *J. Biol. Chem.,* 242, 1651, 1967.
267. **Rao, S. S., Patki, V. M., and Kulkarni, A. D.,** Preparation of active insoluble pepsin, *Indian J. Biochem.,* 7, 210, 1970.
268. **Patel, R. P. and Price, S.,** Derivatives of proteins. I. Polymerization of $\alpha$-chymotrypsin by use of *N*-ethyl-5-phenylisoxazolium-3′-sulfonate, *Biopolymers,* 5, 583, 1967.
269. **Richards, F. M. and Knowles, J. R.,** Glutaraldehyde as a protein cross-linking reagent, *J. Mol. Biol.,* 37, 231, 1968.
270. **Hardy, P. M., Nicholls, A. C., and Rydon, H. N.,** The nature of glutaraldehyde in aqueous solution, *J. Chem. Soc. D,* 10, 565, 1969.
270a. **Habeeb, A. F. S. A. and Hiramoto, R.,** Reactions of proteins with glutaraldehyde, *Arch. Biochem. Biophys.,* 126, 16, 1968.
271. **Ozawa, H.,** Bridging reagent for protein. I. The reaction of diisocyanates with lysine and enzyme proteins, *J. Biochem.,* 62, 419, 1967.
272. **Avrameas, S. and Guilbert, B.,** Biologically active water-insoluble protein polymers. Their use for the isolation of specifically interacting proteins, *Biochimie,* 53, 603, 1971.
273. **Goldman, R. and Katchalski, E.,** Kinetic behavior of a two-enzyme membrane carrying out a consecutive set of reactions, *J. Theor. Biol.,* 32, 243, 1971.
274. **Selegny, E., Broun, G., and Thomas, D.,** Calculation of experimental realization of active transport of neutral molecules in vitro using multilayered, sequential bienzymic, structured membranes, *C. R. Acad. Sci., Ser. D,* 271, 1423, 1970.
275. **Selegny, E., Kernevez, J.-P., Broun, G., and Thomas, D.,** Time-dependent evolutions of diffusion-reactions in model enzymic membranes, *Physiol. Veg.,* 9, 51, 1971.
276. **Bass, L. and McIlroy, D. K.,** Enzyme activities in polarized cell membranes, *Biophys. J.,* 8, 99, 1968.
277. **Schejter, A. and Bar-Eli, A.,** Preparation and properties of crosslinked water-insoluble catalase, *Arch. Biochem. Biophys.,* 136, 325, 1970.
278. **Avrameas, S., Ternynck, T., and Guilbert, B.,** Preparation of insoluble protein derivatives and their utilization in the isolation of antibodies, antigens, enzymes, and enzyme inhibitors, *C. R. Acad. Sci., Ser. D,* 268, 227, 1969.
279. **Manecke, G. and Günzel, G.,** Preparation of water-insoluble active papain, *Naturwissenschaften,* 54, 647, 1967.
280. **Witt, P. R., Jr., Sair, R. A., Richardson, T., and Olson, N. F.,** Chillproofing beer with insoluble papain, *Brew. Dig.,* 45, 70, 1970.
281. **Ashoor, S. H., Sair, R. A., Olson, N. F., and Richardson, T.,** Use of a papain superpolymer to elucidate the structure of bovine casein micelles, *Biochim. Biophys. Acta,* 229, 423, 1971.

282. **Ozawa, H.**, Bridging reagent for protein. II. The reaction of N,N'-polymethylenebis(iodoacetamide) with cysteine and rabbit muscle aldolase, *J. Biochem.*, 62, 531, 1967.

283. **James, L. K. and Augenstein, L.**, Adsorption of enzymes at interfaces: film formation and the effect on activity, *Adv. Enzymol., Relat. Areas Mol. Biol.*, 28, 1, 1966.

284. **Zittle, C. A.**, Adsorption studies of enzymes and other proteins, *Adv. Enzymol., Relat. Areas Mol. Biol.*, 14, 319, 1953.

285. **McLaren, A. D.**, The adsorption and variations of enzymes and proteins on kaolinite, *J. Phys. Chem.*, 58, 129, 1954.

286. **McLaren, A. D., Peterson, G. H., and Barshad, I.**, The adsorption and reactions of enzymes and proteins on clay minerals. IV. Kaolinite and montmorillonite, *Soil Sci. Soc. Am. Proc.*, 22, 239, 1958 (*C. A.*, 52, 20300g, 1958).

287. **Hummel, J. P. and Anderson, B. S.**, Ribonuclease adsorption on glass surfaces, *Arch. Biochem. Biophys.*, 112, 443, 1965.

288. **Messing, R. A.**, Molecular inclusions adsorption of micromolecules on porous glass membranes, *J. Am. Chem. Soc.*, 91, 2370, 1969.

289. **Messing, R. A.**, Immobilized RNase by adsorption on porous glass, *Enzymologia*, 38, 370, 1970.

290. **Mitz, M. A. and Schlueter, R. J.**, Isolation and proteolytic enzymes from solution as dry stable derivatives of cellulosic ion exchanges, *J. Am. Chem. Soc.*, 81, 4024, 1959.

291. **Tosa, T., Mori, T., Fuse, N., and Chibata, I.**, Studies on continuous enzyme reactions. I. Screening of carriers for preparation of water-insoluble aminoacylose, *Enzymologia*, 31, 214, 1966.

292. **Tosa, T., Mori, T., Fuse, N., and Chibata, I.**, Studies on continuous enzyme reactions. II. Preparation of DEAE-cellulose-aminoacylose column and continuous optical resolution of acetyl-DL-methionine, *Enzymologia*, 31, 225, 1966.

293. Anonymous, Enzyme-membrane complexes in module form, *Chem. Eng. News*, 48(50), 39, 1970.

294. **Goldman, R. and Lenhoff, H. M.**, Glucose-6-phosphate dehydrogenase adsorbed on collodion membranes, *Biochim. Biophys. Acta*, 242, 514, 1971.

295. **Thang, M. N., Graffe, M., and Grunberg-Manago, M.**, Observations on the activity of enzymes after filtration on (and through) a mitrocellulose membrane, *Biochem. Biophys. Res. Commun.*, 31, 1, 1968.

296. **Fletcher, G. L. and Okada, S.**, Protection of deoxyribonuclease from ionizing radiation by adsorbents, *Nature*, 176, 882, 1955.

297. **McLaren, A. D. and Estermann, E. F.**, The adsorption and reactions of enzymes and proteins on kaolinite. III. The isolation of enzyme-substrate complexes, *Arch. Biochem. Biophys.*, 61, 158, 1956.

298. **Usami, S., Noda, J., and Goto, K.**, Preparation and properties of water-insoluble saccharase, *Hakko Kagaku Zasshi*, 49, 598, 1971.

299. **Suzuki, H., Ozawa, Y., Maeda, H., and Tanabe, O.**, Water-insoluble enzyme hydrolysis of sucrose by insoluble yeast invertase, *Kogyu Gijutsu-in Hakko Kenkyusho Kenkyu*, No. 31, 11, 1967 (*C. A.*, 67, 113928k, 1967).

300. **Suzuki, H., Ozawa, Y., and Maeda, H.**, Studies on the water-insoluble enzyme hydrolysis of sucrose by insoluble yeast invertase, *Agr. Biol. Chem.*, 30, 807, 1966.

301. **Chung, S.-T., Hamano, M., Aida, K., and Uemura, T.**, Studies on ATP deaminase. Part III. Water insoluble ATP deaminase, *Agr. Biol. Chem.*, 32, 1287, 1968.

302. **Bachler, M. J., Strandberg, G. W., and Smiley, K. L.**, Starch conversion by immobilized glucoamylase, *Biotechnol. Bioeng.*, 12, 85, 1970.

303. **Mitz, M. A.**, New insoluble active derivative of an enzyme as a model for study of cellular metabolism, *Science*, 125, 1076, 1956.

304. **Mandels, M., Kostick, J., and Parizek, R.**, The use of adsorbed cellulase in the continuous conversion of cellulose to glucose, *J. Polym. Sci., Part C*, No. 36, 445, 1971.

305. **Messing, R. A.**, Relationship of pore size and surface area to quantity of stabilized enzyme bound to glass, *Enzymologia*, 39, 12, 1970.

306. **Messing, R. A.**, Insoluble papain prepared by adsorption on porous glass, *Enzymologia*, 38, 39, 1970.

307. **McLaren, A. D.**, Concerning the pH dependence of enzyme reactions on cells, particulates and in solution, *Science*, 125, 697, 1957.

308. **McLaren, A. D. and Estermann, E. F.**, Influence of pH on the activity of chymotrypsin at a solid-liquid interface, *Arch. Biochem. Biophys.*, 68, 157, 1957.

309. **McLaren, A. D.**, Enzyme action in structurally restricted systems, *Enzymologia*, 21, 356, 1960.

310. **Usami, S. and Shirasaki, H.**, Kinetics of enzyme adsorbed on adsorbent, *Hakko Kogaku Zasshi*, 48, 506, 1970.

311. **Tosa, T., Mori, T., Fuse, N., and Chibata, I.**, Studies on continuous enzyme reactions. III. Enzymatic properties of the DEAE-cellulose-aminoacylase complex, *Enzymologia*, 32, 153, 1967.

312. **Sundaram, P. V. and Crook, E. M.**, Preparation and properties of solid-supported urease, *Can. J. Biochem.*, 49, 1388, 1971.

313. **Sundaram, P. V. and Crook, E. M.**, pH-profile of urease adsorbed on kaolinite, *7th Int. Cong. Biochem., Tokyo, Int. Union Biochem.*, 4, Abstr. F-212, 801, 1967.

314. **Tveritinova, E. A., Loboda, N. I., Chukhrai, E. S., and Poltorak, O. M.**, Adsorption and catalytic properties of malate dehydrogenase, *Vestn. Mosk. Univ. Khim.*, 12, 526, 1971 (*C. A.*, 76, 43455p, 1972).

315. **Kimura, A., Shirasaki, H., and Usami, S.**, Kinetics of enzyme adsorbed on adsorbent, *Kogyo Kagaku Zasshi*, 72, 489, 1969.
316. **Tosa, T., Mori, T., and Chibata, I.**, Studies on continuous enzyme reactions. Part VI. Enzymatic properties of the DEAE-Sephadex-aminoacylase complex, *Agr. Biol. Chem.*, 33, 1053, 1969.
317. **Vorobeva, E. S. and Poltorak, O. M.**, Comparative study of enzymic activity of acid phosphatase and phosphoglucomutase in adsorption layers of various types, *Vesin. Mosk. Univ.*, Sec. II 21, 17, 1966 (*C. A.*, 66, 62197n, 1967).
318. **Ogino, S.**, Formation of the fructose-rich polymer by water-insoluble dextransucrase and presence of a glycogen value-lowering factor, *Agr. Biol. Chem.*, 34, 1268, 1970.
319. **Tosa, T., Mori, T., and Chibata, I.**, Studies on continuous enzyme reactions. VII. Activation of water-insoluble aminoacylase by protein denaturing agents, *Enzymologia*, 40, 49, 1971.
320. **Usami, S., Yamada, T., and Kimura, A.**, Activities of enzyme adsorbed on adsorbents. II. Stability of adsorbed enzyme, *Hakko Kyokaishi*, 25, 513, 1967 (*C. A.*, 68, 84499j, 1968).
321. **Gryszkiewicz, J., Dziembor, E., and Ostrowski, W.**, Active insoluble complex of acid phosphomonesterase from the human prostate gland with carboxymethylcellulose, *Bull. Acad. Pol. Sci. Ser. Sci. Biol.*, 18, 439, 1970 (*C. A.*, 74, 38576h, 1971).
322. **Traub, A., Kaufmann, E., and Teitz, Y.**, Synthesis of nicontinamide-adenine dinucleotide by NAD pyrophosphorylase on a column of hydroxylapatite, *Anal. Biochem.*, 28, 469, 1969.
323. **Barth, T. and Maskova, H.**, Continuous resolution of acyl DL-amino acids by carrier-bound kidney acylase. I., *Collect. Czech. Chem. Commun.*, 36, 2398, 1970.
324. **Tosa, T., Mori, T., Fuse, N., and Chibata, I.**, Studies on continuous enzyme reactions. Part V. Kinetics and industrial application of aminoacylase column for continuous optical resolution of acyl-DL-amino acids, *Agr. Biol. Chem.*, 33, 1047, 1969.
325. **Tosa, T., Mori, T., Fuse, N., and Chibata, I.**, Studies on continuous enzyme reactions. IV. Preparation of a DEAE-Sephadex-aminoacylase column and continuous optical resolution of acyl-DL-amino acids, *Biotechnol. Bioeng.*, 9, 603, 1967.
326. **Usami, S. and Taketomi, N.**, Activities of enzyme adsorbed on adsorbents. I., *Hakko Kyokaishi*, 23, 267, 1965 (*C. A.*, 63, 11939f, 1965).
327. **Nikolaev, A. Y. and Mardashev, S. R.**, An insoluble active compound of asparaginase with carboxymethyl cellulose, *Biokhimiya*, 26, 641, 1961 (Engl. trans., 26, 565, 1961).
328. **Kobamoto, N., Löfroth, G., Camp, P., Van Amburg, G., and Augenstein, L.**, Specificity of trypsin adsorption onto cellulose, glass, and quartz, *Biochem. Biophys. Res. Commun.*, 24, 622; 1966.
329. **Negoro, H.**, Continuous inversion of sucrose by using insoluble saccharase, *Hakko Kogaku Zasshi*, 48, 689, 1970.
330. **Koltsova, S. V., Glikina, M. V., and Samsonov, G. V.**, Precipitation of trypsin by poly(methacrylic acid), *Izv. Akad. Nauk. SSSR Ser. Khim.*, 8, 1895, 1970 (*C. A.*, 73, 105665f, 1970).
331. **Harkins, W. D., Fourt, L., and Fourt, P. C.**, Immunochemistry of catalase II. Activity in multilayers, *J. Biol. Chem.*, 132, 111, 1940.
332. **Tveritinova, E. A., Kirai, E., Chukhrai, E. S., and Poltorak, O. M.**, Catalytic activity and adsorption of glucose-6-phosphate dehydrogenase, *Vestn. Mosk. Univ. Khim.*, 24, 16, 1969 (*C. A.*, 73, 73325b, 1970).
333. **Miyamoto, K., Fujii, T., and Miura, Y.**, On the insolubilized enzyme activities using adsorbents and ion exchangers, *Hakko Kogaku Zasshi*, 49, 565, 1971.
334. **Velikanov, L. L., Velikanov, N. L., and Zvyagintsev, D. G.**, Effect of temperature on the activity of free and adsorbed enzymes, *Pochvovedenie*, No. 3, 62, 1971 (*C. A.*, 74, 135651m, 1971).
335. **Goldfeld, M. G., Vorobeva, E. S., and Poltorak, O. M.**, Comparative study of catalase activity in adsorption layers of various types, *Zh. Fiz. Khim.*, 40, 2594, 1966 (*C. A.*, 66, 16713h, 1967).
336. **Langmuir, I. and Schaefer, V. J.**, Properties and structure of protein monolayers, *J. Am. Chem. Soc.*, 61, 181, 1939.
337. **Zhirkov, Y. A., Chukhrai, E. S., and Poltorak, O. M.**, Adsorptive and catalytic activity of hexokinase on silica gel, *Vestn. Mosk. Univ. Khim.*, 12, 405, 1971 (*C. A.*, 76, 43461n, 1972).
338. **Vorobeva, E. S. and Poltorak, O. M.**, Deactivation of hexokinase in adsorption layers of various types, *Zh. Fiz. Khim.*, 40, 2596, 1966 (*C. A.*, 66, 16748y, 1967).
339. **Mkrtumova, N. A. and Deborin, G. A.**, Enzymic activity of ribonuclease adsorbed on sulfonated resin SBS4, *Dokl. Akad. Nauk SSSR*, 146, 1434, 1962 (*C. A.*, 58, 5951a, 1963).
340. **Barnett, L. B. and Bull, H. B.**, The optimum pH of adsorbed ribonuclease, *Biochim. Biophys. Acta*, 36, 244, 1959.
341. **Fletcher, G. and Okada, S.**, Effect of adsorbing materials on radiation inactivation of deoxyribonuclease I, *Radiat. Res.*, 11, 291, 1959.
342. Anonymous, Biochemical engineering finds academic home, *Chem. Eng. News*, 49(26), 23, 1971.
343. **Smiley, K. L.**, Continuous conversion of starch to glucose with immobilized glucoamylase, *Biotechnol. Bioeng.*, 13, 309, 1971.
344. Anonymous, Enzymes on carriers, *Chem. Eng. News*, 49(48), 27, 1971.
345. **Nelson, J. M. and Griffin, E. G.**, Adsorption of invertase, *J. Am. Chem. Soc.*, 38, 1109, 1916.
346. **Griffin, E. G. and Nelson, J. M.**, The influence of certain substances on the activity of invertase, *J. Am. Chem. Soc.*, 38, 722, 1916.

347. **Nelson, J. M. and Hitchcock, D. I.**, The activity of adsorbed invertase, *J. Am. Chem. Soc.*, 43, 1956, 1921.
348. **Langmuir, I. and Schaefer, V. J.**, Activities of urease and pepsin monolayers, *J. Am. Chem. Soc.*, 60, 1351, 1938.
349. **Trurnit, H. J.**, Studies of enzyme systems at a solid-liquid interface I. The system chymotrypsin-serum albumin, *Arch. Biochem. Biophys.*, 47, 251, 1953.
350. **Trurnit, H. J.**, Studies on enzyme systems at a solid-liquid interface II. The kinetics of adsorption and reaction, *Arch. Biochem. Biophys.*, 51, 176, 1954.
351. **Nikolaev, A. Y.**, Catalytic activity of asparaginase in solution and in the adsorbed state, *Biokhimiya*, 27, 843, 1962 (Engl. trans., 27, 713, 1962).
352. **Tosa, T., Mori, T., and Chibata, I.**, Studies on continuous enzyme reactions. VIII. Kinetics and pressure drop of aminoacylase column, *Hakko Kogaku Zasshi*, 49, 522, 1971.
353. **Gale, E. F.**, Studies on bacterial amino-acid decarboxylases. 1. 1(+)-lysine decarboxylase, *Biochem. J.*, 38, 232, 1944.
354. **Becker, W. and Pfeil, E.**, Continuous synthesis of optically active α-hydroxynitriles, *J. Am. Chem. Soc.*, 88, 4299, 1966.
355. **Bernfeld, P. and Wan, J.**, Antigens and enzymes made insoluble by entrapping them into lattices of synthetic polymers, *Science*, 142, 678, 1963.
356. **Dickey, F. H.**, Specific adsorption, *J. Phys. Chem.*, 59, 695, 1955.
357. **Chrambach, A. and Rodbard, D.**, Polyacrylamide gel electrophoresis, *Science*, 172, 440, 1971.
358. **Hicks, G. P. and Updike, S. J.**, The preparation and characterization of lyophilized polyacrylamide enzyme gels for chemical analysis, *Anal. Chem.*, 38, 726, 1966.
359. **van Duijn, P., Pascoe, E., and van der Ploeg, M.**, Theoretical and experimental aspects of enzyme determination in a cytochemical model system of polyacrylamide films containing alkaline phosphatase, *J. Histochem. Cytochem.*, 15, 631, 1967.
360. **Bernfeld, P., Bieber, R. E., and MacDonnell, P. C.**, Water-insoluble enzymes: arrangement of aldolase within an insoluble carrier, *Arch. Biochem. Biophys.*, 127, 779, 1968.
361. **Mosbach, K. and Mosbach, R.**, Entrapment of enzymes and microorganisms in synthetic cross-linked polymers and their application in column techniques, *Acta Chem. Scand.*, 20, 2807, 1966.
362. **Brown, H. D., Patel, A. B., and Chattopadhyay, S. K.**, Enzyme entrapment within hydrophobic and hydrophilic matrices, *J. Biomed. Mater. Res.*, 2, 231, 1968.
363. **Brown, H. D., Patel, A. B., and Chattopadhyay, S. K.**, Lattice entrapment of glycolytic enzymes, *J. Chromatogr.*, 35, 101, 1968.
364. **Degani, Y. and Miron, T.**, Immobilization of cholinesterase in cross-linked polyacrylamide, *Biochim. Biophys. Acta*, 212, 362, 1970.
365. **Dobo, J.**, Application of radiation polymerization in the production of water-insoluble enzyme preparations, *Acta Chim. Acad. Sci. Hung.*, (Budapest), 63, 453, 1970.
366. **Strandberg, G. W. and Smiley, K. L.**, Free and immobilized glucose isomerase from *Streptomyces phaeochromogenes*, *Appl. Microbiol.*, 21, 588, 1971.
367. **Bernfeld, P., Bieber, R. E., and MacDonnell, P. C.**, Water-insoluble enzymes: $^{14}$C-labeled, crystallized aldolase, *Fed. Proc.*, 27, 782, 1968.
368. **Pennington, S. N., Brown, H. D., Patel, A. B., and Knowles, C. O.**, Properties of matrix supported acetylcholinesterase, *Biochim. Biophys. Acta*, 167, 476, 1968.
369. **Pennington, S. N., Brown, H. D., Patel, A. B., and Chattopadhyay, S. K.**, Silastic entrapment of glucose-peroxidase and acetylcholinesterase, *J. Biomed. Mater. Res.*, 2, 443, 1968.
370. **Guilbault, G. G. and Das, J.**, Immobilization of cholinesterase and urease, *Anal. Biochem.*, 33, 341, 1970.
371. **Johnson, P. and Whateley, T. L.**, On the use of polymerizing silica gel systems for the immobilization of trypsin, *J. Colloid Interface Sci.*, 37, 557, 1971.
372. **Bauman, E. K., Goodson, L. H., Guilbault, G. G., and Kramer, D. N.**, Preparation of immobilized cholinesterase for use in analytical chemistry, *Anal. Chem.*, 37, 1378, 1965.
373. **Guilbault, G. G. and Kramer, D. N.**, Fluorometric system employing immobilized cholinesterase for assaying anticholinesterase compounds, *Anal. Chem.*, 37, 1675, 1965.
374. **Bauman, E. K., Goodson, L. H., and Thomson, J. R.**, Stabilization of serum cholinesterase in dried starch gel, *Anal. Biochem.*, 19, 587, 1967.
375. SNAM Progetti S.p.A., Enzyme-containing fibers for catalysis of enzymic reactions, Ger. 1,932,426, Jan. 2, 1970.
376. SNAM Progetti S.p.A., Improvements in or relating to structures containing enzyme material, Brit. 1,224,947, March 10, 1971.
377. Anonymous, Fibers trap enzymes, *Chem. Eng. News*, 49(16), 56, 1971.
377a. Anonymous, E.N.I. traps enzymes, *Eur. Chem. News*, 19(480), 31, 1971.
378. **Wieland, T., Determann, H., and Buennig, K.**, Insoluble enzymes fixed in polyacrylamide gel, *Z. Naturforsch. b.*, 21, 1003, 1966.
379. **Mosbach, K. and Larsson, P.-O.**, Preparation and application of polymer-entrapped enzymes and microorganisms in microbial transportation processes with special reference to steroid 11-β-hydroxylation and $\Delta^1$-dehydrogenation, *Biotechnol. Bioeng.*, 12, 19, 1970.

380. **Wingard, L. B., Jr., Liu, C. C., and Nagda, N. L.,** Electrochemical measurements with glucose oxidase immobilized in polyacrylamide gel: constant current voltametry, *Biotechnol. Bioeng.,* 13, 629, 1971.
381. **Bernfeld, P., Bieber, R. E., and Watson, D. M.,** Kinetics of water-insoluble phosphoglycerate mutase, *Biochim. Biophys. Acta,* 191, 570, 1969.
382. **Bernfeld, P. and Bieber, R. E.,** Water-insoluble enzymes: kinetics of rabbit muscle enolase embedded within an insoluble carrier, *Arch. Biochem. Biophys.,* 131, 587, 1969.
383. **Guilbault, G. G. and Hrabankova, E.,** An electrode for determination of amino acids, *Anal. Chem.,* 42, 1779, 1970.
384. **Guilbault, G. G. and Montalvo, J. G., Jr.,** Improved urea-specific enzyme electrode, *Anal. Lett.,* 2, 283, 1969.
385. **Guilbault, G. G. and Montalvo, J. G.,** An enzyme electrode for the substrate urea, *J. Am. Chem. Soc.,* 92, 2533, 1970.
386. **Updike, S. J. and Hicks, G. P.,** Reagentless substrate analysis with immobilized enzymes, *Science,* 158, 270, 1967.
387. **Updike, S. J. and Hicks, G. P.,** The enzyme electrode, *Nature,* 214, 986, 1967.
388. **Guilbault, G. G. and Hrabankova, E.,** New enzyme electrode probes for D-amino acids and asparagine, *Anal. Chim. Acta,* 56, 285, 1971.
389. **van der Ploeg, M. and van Duijn, P.,** 5,6-Dihydroxy indole as a substrate in a histochemical peroxidase reaction, *J. Roy. Microscop. Soc.,* 83, 415, 1964.
390. **Guilbault, G. G. and Shu, F. R.,** An electrode for the determination of glutamine, *Anal. Chim. Acta,* 56, 333, 1971.
391. **Guilbault, G. G. and Montalvo, J. G., Jr.,** A urea-specific enzyme electrode, *J. Am. Chem. Soc.,* 91, 2164, 1969.
392. **Montalvo, J. G., Jr. and Guilbault, G. G.,** Sensitized cation selective electrode, *Anal. Chem.,* 41, 1897, 1969.
393. **Guilbault, G. G. and Hrabankova, E.,** Determination of urea in blood and urine with a urea-sensitive electrode, *Anal. Chim. Acta,* 52, 287, 1970.
394. **Chang, T. M. S.,** Semipermeable microcapsules, *Science,* 146, 524, 1964.
395. **Chang, T. M. S. and MacIntosh, F. C.,** Semipermeable aqueous microcapsules, *Pharmacologist,* 6, 198, 1964.
396. **Chang, T. M. S.,** Semipermeable aqueous microcapsules, Thesis, McGill University, Montreal, Canada, 1965.
397. **Herbig, J. A.,** Microencapsulation, *Encyl. Polym. Sci. Technol.,* 8, 719, 1968.
398. **Chang, T. M. S., MacIntosh, F. C., and Mason, S. G.,** Semipermeable aqueous microcapsules. I. Preparation and properties, *Can. J. Physiol. Pharmacol.,* 44, 115, 1966.
399. **Chang, T. M. S., Johnson, L. J., and Ransome, O. J.,** Semipermeable aqueous microcapsules. IV. Nonthrombogenic microcapsules with heparin-complexed membranes, *Can. J. Physiol. Pharmacol.,* 45, 705, 1967.
400. **Chang, T. M. S.,** Microcapsules as artifical cells, *Sci. J.,* 3, 62, 1967.
401. **Chang, T. M. S.,** Semipermeable aqueous microcapsules ("artificial cells"): with emphasis on experiments in an extracorporeal shunt system, *Trans. Am. Soc. Artif. Intern. Organs,* 12, 13, 1966.
402. **Chang, T. M. S. and Poznansky, M. J.,** Semipermeable microcapsules containing catalase for enzyme replacement in acatalasaemic mice, *Nature,* 218, 243, 1968.
403. **Chang, T. M. S., Gonda, A., Dirks, J. H., and Malave, N.,** Clinical evaluation of chronic, intermittent, and short term hemoperfusions in patients with chronic renal failure using semipermeable microcapsules (artificial cells) formed from membrane-coated activated charcoal, *Trans. Am. Soc. Artif. Intern. Organs,* 17, 246, 1971.
404. **Chang, T. M. S., Pont, A., Johnson, L. J., and Malave, N.,** Response to intermittent extracorporeal perfusion through shunts containing semipermeable microcapsules, *Trans. Am. Soc. Artif. Intern. Organs,* 14, 163, 1968.
405. **Kitajima, M., Miyano, S., and Kondo, A.,** Enzyme-containing microcapsules, *Kogyo Kagaku Zasshi,* 72, 493, 1969 (*C. A.,* 70, 118067a, 1969).
406. **Kitajima, M. and Kondo, A.,** Fermentation without multiplication of cells using microcapsules that contain zymase complex and muscle enzyme extract, *Bull. Chem. Soc. Jap.,* 44, 3201, 1971.
407. **Levine, S. N. and LaCourse, W. C.,** Materials and design consideration for a compact artificial kidney, *J. Biomed. Mater. Res.,* 1, 275, 1967.
408. **Sparks, R. E., Salemme, R. M., Meier, P. M., Litt, M. H., and Lindan, O.,** Removal of waste metabolites in uremia by microencapsulated reactants, *Trans. Am. Soc. Artif. Intern. Organs,* 15, 353, 1969.
409. **Boguslaski, R. C. and Janik, A. M.,** A kinetic study of microencapsulated bovine carbonic anhydrase, *Biochim. Biophys. Acta,* 250, 266, 1971.
410. **Chang, T. M. S.,** The in vivo effects of semipermeable microcapsules containing L-asparaginase on 6C3HED lymphosarcoma, *Nature,* 229, 117, 1971.
411. **Morgan, P. W. and Kwolek, S. L.,** Interfacial polycondensation. II. Fundamentals of polymer formation at liquid interfaces, *J. Polym. Sci.,* 40, 299, 1959.
412. **Shiba, M., Tomioka, S., Koishi, M., and Kondo, T.,** Studies on microcapsules. V. Preparation of polyamide microcapsules containing aqueous protein solution, *Chem. Pharm. Bull. (Tokyo),* 18, 803, 1970.
413. **Chang, T. M. S. and Poznansky, M. J.,** Semipermeable aqueous microcapsules (artificial cells). V. Permeability characteristics, *J. Biomed. Mater. Res.,* 2, 187, 1968.
414. **Chang, T. M. S.,** Artificial cells made to order, *New Sci.,* 42, 18, 1969.
415. **Chang, T. M. S. and Malave, N.,** The development and first clinical use of semipermeable microcapsules (artificial cells) as a compact artificial kidney, *Trans. Am. Soc. Artif. Intern. Organs,* 16, 141, 1970.
416. **Esso Research and Engineering Co.,** Separating hydrocarbons with liquid membranes, U.S. 3,410,794, Nov. 12, 1968.

417. **Li, N. N.**, Separation of hydrocarbons by liquid membrane permeation, *Ind. Eng. Chem. Process Des. Develop.*, 10, 215, 1971.
418. **Li, N. N.**, Permeation through liquid surfactant membranes, *AIChE J.*, 17, 459, 1971.
419. **Li, N. N. and Shrier, A. L.**, Liquid membrane water treating, in *Recent Developments in Separation Science*, Vol. 1, Li, N. N., Ed., Chemical Rubber Co., Cleveland, 1972, 163.
420. **Shah, N. D. and Owens, T. C.**, Separation of benzene and hexane with liquid membrane technique, *Ind. Eng. Chem. Prod. Res. Develop.*, 11, 58, 1972.
421. **May, S. W. and Li, N. N.**, The immobilization of urease using liquid-surfactant membranes, *Biochem. Biophys. Res. Commun.*, 47, 1179, 1972.
422. **Li, N. N., Brusca, D. R., and Mohan, R. R.**, personal communication.
423. **Mohan, R. R. and Li, N. N.**, personal communication.
424. **van Oss, C. J.**, Ultrafiltration membranes, in *Progress in Separation and Purification*, Vol. 3, Perry, E. S. and van Oss, C. J., Eds., John Wiley & Sons, New York, 1970, 97.
425. **Michaels, A. S.**, Ultrafiltration, in *Progress in Separation and Purification*, Vol. 1, Perry, E. S., Ed., John Wiley & Sons, New York, 1968, 297.
426. **Michaels, A. S. and Bixler, H. J.**, Membrane permeation: theory and practice, in *Progress in Separation and Purification*, Vol. 1, Perry, E. S., Ed., John Wiley & Sons, New York, 1968, 143.
427. **Wang, D. I. C. and Humphrey, A. E.**, Biochemical engineering, *Chem. Eng.*, 76, 108, 1969.
428. **Butterworth, T. A., Wang, D. I. C., and Sinskey, A. J.**, Application of ultrafiltration for enzyme retention during continuous enzymatic reaction, *Abstr. 158th Meet. Am. Chem. Soc., Micr.*, Contribution 5, 1969.
429. **Butterworth, T. A., Wang, D. I. C., and Sinskey, A. J.**, Application of ultrafiltration for enzyme retention during continuous enzymatic reaction, *Biotechnol. Bioeng.*, 12, 615, 1970.
430. **Ghose, T. K. and Kostick, J. A.**, A model for continuous enzymatic saccharification of cellulose with simultaneous removal of glucose syrup, *Abstr. 158th Meet. Am. Chem. Soc., Micr.*, Contribution 4, 1969.
431. **Ghose, T. K. and Kostick, J. A.**, A model for continuous enzymatic saccharification of cellulose with simultaneous removal of glucose syrup, *Biotechnol. Bioeng.*, 12, 921, 1970.
432. **Marshall, J. J. and Whelan, W. J.**, A new approach to the use of enzymes in starch technology, *Chem. Ind.*, No. 25, 701, 1971.
433. **Stavenger, P. L.**, Putting semipermeable membranes to work, *Chem. Eng. Progr.*, 67, 30, 1971.
434. **Rony, P. R.**, Multiphase catalysis. II. Hollow fiber catalysts, *Biotechnol. Bioeng.*, 13, 431, 1971.
435. **Rony, P. R.**, personal communication.
436. **Williams, D. L., Doig, A. R., Jr., and Korosi, A.**, Electrochemical enzymatic analysis of blood glucose and lactate, *Anal. Chem.*, 42, 118, 1970.
437. **Guilbault, G. G. and Hrabankova, E.**, L-Amino acid electrode, *Anal. Lett.*, 3, 53, 1970.
438. **Clark, L. C., Jr. and Lyons, C.**, Electrode systems for continuous monitoring in cardiovascular surgery, *Ann. NY Acad. Sci.*, 102, 29, 1962.
439. **Blumenthal, R., Caplan, S. R., and Kedem, O.**, The coupling of an enzymatic reaction to transmembrane flow of electric current in a synthetic "active transport" system, *Biophys. J.*, 7, 735, 1967.
440. **Lilly, M. D. and Dunnill, P.**, Biochemical reactors, *Process Biochem.*, 6, 29, 1971.
441. **Sundaram, P. V., Tweedale, A., and Laidler, K. J.**, Kinetic laws for solid-supported-enzymes, *Can. J. Chem.*, 48, 1498, 1970.
442. **O'Neill, S. P., Lilly, M. D., and Rowe, P. N.**, Multiple steady states in stirred tank enzyme reactors, *Chem. Eng. Sci.*, 26, 173, 1971.
443. **Lilly, M. D., Kay, G., Sharp, A. K., and Wilson, R. J. H.**, The operation of biochemical reactors using fixed enzymes, *Biochem. J.*, 5p, 1968.
444. **Kobayashi, T. and Moo-Young, M.**, Backmixing and mass transfer in the design of immobilized-enzyme reactors, *Biotechnol. Bioeng.*, 13, 893, 1971.
445. **Reese, E. T. and Mandels, M.**, Enzyme action on partition chromatographic columns, *J. Am. Chem. Soc.*, 80, 4625, 1958.
446. **Guilbault, G. G.**, Use of enzymes in analytical chemistry, *Anal. Chem.*, 38, 527R, 1966.
447. **Guilbault, G. G.**, Use of enzymes in analytical chemistry, *Anal. Chem.*, 40, 459R, 1968.
448. **Guilbault, G. G.**, Enzymic methods of analysis, *Rec. Chem. Progress*, 30, 261, 1969.
449. **Guilbault, G. G.**, Enzymic methods of analysis, *CRC Crit. Rev. Anal. Chem.*, 1, 377, 1970.
450. **Durst, R. A.**, Ion-selective electrodes in science, medicine, and technology, *Am. Sci.*, 59, 353, 1971.
451. **Young, T. G., Hadjipetrou, L., and Lilly, M. D.**, The theoretical aspects of biochemical fuel cells, *Biotechnol. Bioeng.*, 8, 581, 1966.
452. **Larsson, P.-O. and Mosbach, K.**, Preparation of a NAD(H)-polymer matrix showing coenzymic function of the bound pyridine nucleotide, *Biotechnol. Bioeng.*, 13, 393, 1971.
453. **Weibel, M. K., Weetall, H. H., and Bright, H. J.**, Insolubilized coenzymes: the covalent coupling of enzymatically active NAD to glass surfaces, *Biochem. Biophys. Res. Commun.*, 44, 347, 1971.
454. **Lowe, C. R. and Dean, P. D. G.**, Affinity chromatography of enzymes on insolubilized cofactors, *FEBS Lett.*, 14, 313, 1971.

455. **Mosbach, K., Guilford, H., Larsson, P.-O., Ohlsson, R., and Scott, M.,** Purification of nicotinamide-adenine dinucleotide-dependent dehydrogenases by affinity chromatography, *Biochem. J.,* 125, 20p, 1971.
456. **Wilchek, M., Salomon, Y., Lowe, M., and Selinger, Z.,** Conversion of protein kinase to a cyclic AMP independent form by affinity chromatography on $N^6$-caproyl 3′,5′-cyclic adenosine monophosphate-Sepharose, *Biochem. Biophys. Res. Commun.,* 45, 1177, 1971.
457. **Harpold, M. A. and Calvin, M.,** AMP on an insoluble solid support, *Nature,* 219, 486, 1968.
458. **Penzer, G. R. and Radda, G. K.,** Flavins in a solid matrix, *Nature,* 213, 251, 1967.
459. **Grubbs, R. H. and Kroll, L. C.,** Catalytic reduction of olefins with a polymer-supported rhodium(I) catalyst, *J. Am. Chem. Soc.,* 93, 3062, 1971.
460. **Grubbs, R. H. and Kroll, L. C.,** The selectivity of polymer supported homogeneous catalysts, *Abstr. 162nd Meet. Am. Chem. Soc., Polymer Chem.,* Contribution 68, 1971.
461. **Mobil Oil Corp.,** Improvements relating to organic reactions catalyzed by resin-metal complexes, Brit. 1,238,703, July 7, 1971.
462. **British Petroleum Co., Ltd.,** New catalyst supports and catalysts contained in these new supports, Ger. 2,062,351, June 24, 1971.
463. **British Petroleum Co., Ltd.,** Method for hydroformylation of olefins, Ger., 2,062,352, June 24, 1971.
464. **Feinstein, G.,** Purification of trypsin by affinity chromatography on ovomucoid-Sepharose resin, *FEBS Lett.,* 7, 353, 1970.
465. **Kasche, V.,** Activity distribution in modified α-chymotrypsin, *Biochem. Biophys. Res. Commun.,* 38, 875, 1970.
466. **Robinson, N. C., Tye, R. W., Neurath, H., and Walsh, K. A.,** Isolation of trypsins by affinity chromatography, *Biochemistry,* 10, 2743, 1971.
467. **Krug, F., Desbuquois, B., and Cuatrecasas, P.,** Glucagon affinity adsorbent: selective binding of acceptors of liver cell membranes, *Nature, New Biol.,* 234, 268, 1971.
468. **Richardson, M. C. and Schulster, D.,** β1-24-Adrenocorticotrophin diazotized to polyacrylamide: effects on isolated adrenal cells, *Biochem. J.,* 125, 60p, 1971.
469. **Selinger, R. C. L. and Civen, M.,** ACTH diazotized to agarose: effects on isolated adrenal cells, *Biochem. Biophys. Res. Commun.,* 43, 793, 1971.
470. **Donnelly, E. H. and Goldstein, I. J.,** Glutaraldehyde-insolubilized concanavalin A: an adsorbent for the specific isolation of polysaccharides and glycoproteins, *Biochem. J.,* 118, 679, 1970.
471. **Poonian, M. S., Schlabach, A. J., and Weissbach, A.,** Covalent attachment of nucleic acids to agarose for affinity chromatography, *Biochemistry,* 10, 424, 1971.
472. **Wagner, A. F., Bugianesi, R. L., and Shen, T. Y.,** Preparation of Sepharose-bound poly (rI:rC), *Biochem. Biophys. Res. Commun.,* 45, 184, 1971.
473. **Iverius, P.-H.,** Coupling of glycosaminoglycans to agarose beads (Sepharose 4B), *Biochem. J.,* 124, 677, 1971.
474. **Manecke, G. and Lamer, W.,** Separation of racemates on optically active polymers, *Naturwissenschaften,* 54, 647, 1967.
475. **Saxinger, C., Ponnamperuma, C., and Woese, C.,** Evidence for the interaction of nucleotides with immobilized amino acids and its significance for the origin of the genetic role, *Nature, New Biol.,* 234, 172, 1971.
476. **Olesen, H., Hippe, E., and Haber, E.,** Nature of vitamin $B_{12}$ binding. I. Covalent coupling of hydroxocobalamin to soluble and insoluble carriers, *Biochim. Biophys. Acta,* 243, 66, 1971.
477. **Cuatrecasas, P. and Anfinsen, C. B.,** Affinity chromatography, in *Methods Enzymol.,* Vol. 22, Jakoby, W. B., Ed., Academic Press, New York, 1971, 345.
478. **Cuatrecasas, P.,** Selective adsorbents based on biochemical specificity, in *Biochemical Aspects of Reactions on Solid Supports,* Stark, G. R., Ed., Academic Press, New York, 1971, 79.
479. **Lerman, L. S.,** Biochemically specific method for enzyme isolation, *Proc. Natl. Acad. Sci. USA,* 39, 232, 1953.
480. **Marshall, G. R. and Merrifield, R. B.,** Solid phase synthesis: the use of solid supports and insoluble reagents in peptide synthesis, in *Biochemical Aspects of Reactions on Solid Supports,* Stark, G. R., Ed., Academic Press, New York, 1971, 111.
481. **Stark, G. R.,** Sequential degradation of peptides using solid supports, in *Biochemical Aspects of Reactions on Solid Supports,* Stark, G. R., Ed., Academic Press, New York, 1971, 171.
482. **Laursen, R. A.,** Solid-phase Edman degradation on automatic peptide sequences, *Eur. J. Biochem.,* 20, 89, 1971.

**Government Reports**

**Drake, R. F., Deibert, M. C., Matsuda, S., O'Connell, J. J., and Nuwayser, E. S.,** Implantable fuel cell for an artificial heart, *U.S. Clearinghouse Fed. Sci. Tech. Inform., PB Rep.,* 1968, PB 177695 (*C. A.,* 69, 112748f, 1968).

**Lynn, J., Emmerling, D., and Falb, R. D.,** Studies on optimization of techniques for enzyme insolubilization, *U.S. Natl. Tech. Inform. Serv.,* NASA CR-73354 (N70-23428), 1969.

**Slote, L.,** Development of immobilized enzyme systems for enhancement of biological waste treatment processes, *U.S. Natl. Tech. Inform. Serv., PB Rep.,* 1970, No. 203598 (*C. A.,* 76, 76173y, 1972).

**Patents**

Wallerstein Co., Inc., Dextrose preparation by use of starch-glucogenase enzymes, U.S. 2,717,852, Sept. 13, 1955.

Armour-Pharmaceutical Co., Catalase compositions, U.S. 3,126,324, March 24, 1964.

Leuschner, F., Shaped structures for biological processes, Brit. 953,414, March 25, 1964.

Melpar, Inc., Immobilized serum cholinesterase, U.S. 3,223,593, Dec. 14, 1965.

Miles Laboratories, Inc., Diagnostic compositions and test indicators, U.S. 3,235,337, Feb. 15, 1966.

Yeda Research & Development Co., Water insoluble enzymes, U.S. 3,278,392, Oct. 11, 1966.

Monsanto Company, Method for preparing insoluble enzymes and protein, Netherlands 6,708,738, Dec. 27, 1967.

Kyowa Hakko Kogyo Kabushiki Kaisha, Water-insoluble enzymes, Brit. 1,108,533, April 3, 1968.

National Research Development Corp., Novel polymeric materials, S. African 68 4165, June 25, 1968.

National Research Development Corp., Polymers having dihalo-*s*-triazinyl groups bound thereto, S. African 68 04,009, Nov. 12, 1968.

National Research Development Corp., Polymeric matrixes having biologically active substances bound thereto, S. African 68 04,010, Nov. 12, 1968.

National Research Development Corp., Polymers having triazinyl groups capable of acting as supports for biologically active substrates, S. African 68 04,164, Nov. 14, 1968.

National Research Development Corp., Polymeric membranes containing enzymes, S. African 68 04,165, Nov. 14, 1968.

Max-Planck-Gesellschaft zur Förderung der Wissenschaften e.V., High molecular weight polymers for binding enzymes, Ger. 1,282,579, Nov. 14, 1968.

Monsanto Company, Insoluble enzymes and proteins, Fr. 1,559,606, March 14, 1969.

Corning Glass Works, Stabilization of enzymes by adsorption of an inorganic support, Fr. 2,001,336, Sept. 26, 1969.

Agence Nationale de Valorisation de la Recherche, Method of fixing active proteins on a carrier, Ger. 1,915,970, Oct. 9, 1969.

National Research Development Corp., Production of polymeric matrices having biologically active substances chemically bound to them, Brit. 1,183,257, March 4, 1970.

National Research Development Corp., Production of a polymer having dihalo-*s*-triazinyl groups bound to it for use as a polymeric matrix for biologically active substances, Brit. 1,183,258, March 4, 1970.

National Research Development Corp., Polymeric materials carrying substantial triazinyl groups capable of acting as a support for biologically active substances, Brit. 1,183,259, March 4, 1970.

National Research Development Corp., Insolubilized enzymes, their preparation and use in enzymatic reactions, Brit. 1,183,260, March 4, 1970.

Tanabe Seiyaku Co., Ltd., Aspartase on anion exchangers, Japan 70 06,870, March 9, 1970.

Teijin Ltd., Enzymic resins, Japan 70 06,875, March 9, 1970.

Unilever N. V., Enzymically active materials, Ger. 1,939,347, March 26, 1970.

Monsanto Company, Valence-bonded enzymically active polymer adducts, Ger. 1,943,490, April 2, 1970.

Monsanto Company, Enzymically active adducts of asparaginase and maleic anhydride copolymers, Ger. 1,945, 748, April 2, 1970.

Yeda Research and Development Co., Valence-bonded enzymically active polymer adducts, Ger. 1,944,369, April 2, 1970.

Monsanto Company, Enzymic treatment of a substrate using enzyme-polymer compounds, Ger. 1,948,273, April 16, 1970.

Muanyagipari Kutato Intezet, Highly active and stable enzyme and antigen preparations insoluble in water, Hung. Teljes 321, April 22, 1970.

Corning Glass Works, Stabilized insoluble enzyme preparations, Ger. 1,944,418, April 23, 1970.

Corning Glass Works, Water-insoluble stabilized alkaline protease and glucose oxidase, Ger. 1,949,943, April 30, 1970.

Corning Glass Works, Immobilized water-insoluble hydrolases and oxidoreductases, Ger. 1,905,681, May 21, 1970.

National Research Development Corp., Enzymic reactions with insoluble enzymes, Ger. 1,959,169, June 4, 1970.

Intermag. A.-G., Enzymic treatment of liquids, Ger. 1,809,830, June 11, 1970.

Maunyagipari Kutato Intezet, Stabilized enzyme and antigen preparations, Ger. 1,955,638, June 11, 1970.

Tanabe Seiyaku Co., Ltd., Water-insoluble aspartase preparations, Japan 70 17,587, June 17, 1970.

Ranks Hovis McDougall Ltd., Water-insoluble amylases coupled with diazotized cellulose ethers, Ger. 1,953,189, June 18, 1970.

Monsanto Company, Detergent containing soluble polymer enzyme products, Ger. 1,948,177, July 2, 1970.

Corning Glass Works, Chemically coupled enzymes, U.S. 3,519,538, July 7, 1970.

Boehringer Mannheim G. m. b. H., Medicinal disc-shaped particles containing enzymes, Brit. 1,202,050, August 12, 1970.

Monsanto Company, Soluble enzyme-polymer adducts, Ger. 1,945,680, Aug. 13, 1970.

Boehringer Mannheim G. m. b. H., Copolymers as carriers for proteins insoluble in water, Ger. 1,935,711, August 20, 1970.

Ranks Hovis McDougall, Ltd., (*p*-Diazophenoxy)hydroxypropyl cellulose as carrier for enzymes, Ger. 2,012,089, Sept. 24, 1970.

Procter and Gamble Co., Protease containing enzymatic compositions with polymethacryloyl chloride resins matrix, Ger. 2,047,495, Sept. 26, 1970.

A. Guinness Son & Co., Insoluble active enzymes, Netherlands 7,004,923, Oct. 13, 1970.

A. Guinness Son & Co., Enzymic compositions, Ger. 2,016,729, Oct. 15, 1970.

Teijin Ltd., Water-insoluble enzyme resins, U.S. 3,536,587, Oct. 27, 1970.

Kyowa Fermentation Industry Co., Ltd., Continuous enzyme reactions, Ger. 2,008,842, Nov. 5, 1970.

Koch-Light Laboratories Ltd., Hydrophilic water-insoluble copolymers, Ger. 2,014,122, Nov. 12, 1970.

E. I. du Pont de Nemours & Co., Enzyme electrode, U.S. 3,542,662, Nov. 24, 1970.

Monsanto Company, Stable, mouth hygiene agent containing a polymer-enzyme compound, Ger. 1,948,298, Nov. 26, 1970.

Beckman Instruments, Inc., Immobilizing enzymes for storage purposes by coupling to proteins, Brit. 1,216,512, Dec. 23, 1970.

CPC International Inc., Separation of enzymes from liquid media, Ger. 2,039,222, Feb. 25, 1971.

Procter and Gamble Co., Water-insoluble protease resin adducts for detergents, Ger. 2,047,495, April 8, 1971.

Procter and Gamble Co., Protease-containing enzymatic compositions with diazotized polyanhydroglucose anthranilate matrix, Ger. 2,047,496, April 8, 1971.

Pharmacia AB, Fixing enzymes antibodies and antigenic proteins to polymers, Switzerland 503 707, April 14, 1971.

Baxter Laboratories, Inc., Streptokinase binding to carbohydrates, Ger. 2,059,165, June 9, 1971.

Worthington Biochemical Corp., Enzymic preparations, Ger. 2,060,121, June 16, 1971.

Exploaterings Aktiebolag T. B. F., Fixed polymeric products, especially adsorption materials and polymer-bound enzymes, Ger. 2,061,009, June 24, 1971.

Ranks Hovis McDougall Ltd., Water-insoluble enzyme-cellulose carbonate coupling product, Ger. 2,062,246, July 1, 1971.

Baxter Laboratories, Inc., Immobilized glucose isomerase-basic anion exchanger cellulose complex, Ger. 2,061,371, July 1, 1971.

Monsanto Company, Stable agglomerated enzyme products for use in detergents, U.S. 3,594,325, July 20, 1971.

Exploaterings Aktiebolag T. B. F., Coupling of trypsin or trypsin inhibitor with epoxide-containing polymers, Ger. 2,102,514, July 29, 1971.

American Cyanamid Co., Enzyme stabilization, Ger. 2,104,810, August 19, 1971.

Monsanto Company, Insoluble enzymic compositions, Ger. 2,106,213, August 19, 1971.

Monsanto Company, Insoluble enzymic compositions, Ger. 2,106,214, August 19, 1971.

Monsanto Company, Silicon-containing material on whose surface is attached an enzyme-polymer product, Ger. 2,106,215, August 19, 1971.

Merck Patent G. m. b. H., Maleic anhydride copolymers as carriers for reactants, Ger. 2,008,990, Sept. 9, 1971.

Monsanto Company, Reactors containing enzymes for adsorption to polymeric surfaces, Ger. 2,112,740, Oct. 21, 1971.

# INDEX

## A

## B



## C

## D

## E



## F



## G

H



I



K



L

M



N



O

## S



## T

## U



## V



## W



## X



## Y



## Z